NJU SA 2020-2021

南京大学建筑与城市规划学院 建筑系教学年鉴
THE YEAR BOOK OF ARCHITECTURE DEPARTMENT TEACHING PROGRAM
SCHOOL OF ARCHITECTURE AND URBAN PLANNING NANJING UNIVERSITY
胡友培 编 EDITOR : HU YOUPEI
东南大学出版社·南京 SOUTHEAST UNIVERSITY PRESS, NANJING

建筑设计及其理论
Architectural Design and Theory

张 雷 教 授	Professor ZHANG Lei
冯金龙 教 授	Professor FENG Jinlong
吉国华 教 授	Professor JI Guohua
周 凌 教 授	Professor ZHOU Ling
傅 筱 教 授	Professor FU Xiao
王 铠 副研究员	Associate Researcher WANG Kai
钟华颖 副研究员	Associate Researcher ZHONG Huaying
黄华青 副研究员	Associate Researcher HUANG Huaqing
梁宇舒 助理研究员	Asistant Researcher LIANG Yushu

城市设计及其理论
Urban Design and Theory

丁沃沃 教 授	Professor DING Wowo
鲁安东 教 授	Professor LU Andong
华晓宁 副教授	Associate Professor HUA Xiaoning
胡友培 副教授	Associate Professor HU Youpei
窦平平 副教授	Associate Professor DOU Pingping
刘 铨 副教授	Associate Professor LIU Quan
尹 航 讲 师	Lecturer YIN Hang
唐 莲 副研究员	Associate Researcher TANG lian
尤 伟 副研究员	Associate Researcher YOU Wei

建筑历史与理论及历史建筑保护
Architectural History and Theory, Protection of Historic Buildings

赵 辰 教 授	Professor ZHAO Chen
王骏阳 教 授	Professor WANG Junyang
胡 恒 教 授	Professor HU Heng
冷 天 副教授	Associate Professor LENG Tian
史文娟 副研究员	Associate Researcher SHI Wenjuan

建筑技术科学
Building Technology Science

吴 蔚 副教授	Associate Professor WU Wei
郜 志 副教授	Associate Professor GAO Zhi
童滋雨 副教授	Associate Professor TONG Ziyu
梁卫辉 副教授	Associate Professor LIANG Weihui
施珊珊 副教授	Associate Professor SHI Shanshan
孟宪川 副研究员	Associate Researcher MENG Xianchuan
李清朋 副研究员	Associate Researcher LI Qingpeng
王力凯 博士后	Postdoctor WANG Likai

南京大学建筑与城市规划学院建筑系
Department of Architecture
School of Architecture and Urban Planning
Nanjing University
arch@nju.edu.cn　　http://arch.nju.edu.cn

教学纲要
EDUCATIONAL PROGRAM

教学阶段 Phases of Education	本科生培养（学士学位）Undergraduate Program (Bachelor Degree)			
	一年级 1st Year	二年级 2nd Year	三年级 3rd Year	四年级 4th Year
教学类型 Types of Education	通识教育 General Education			专业教育 Professional Education
课程类型 Types of Courses	通识类课程 General Courses	学科类课程 Disciplinary Courses		专业类课程 Professional Courses
主干课程 Design Courses	设计基础 Design Foundation	建筑设计基础 Architectural Design Foundation	建筑设计 Architectural Design	
理论课程 Theoretical Courses	基础理论 Basic Theory of Architecture		专业理论 Architectural Theory	
技术课程 Technological Courses				
实践课程 Practical Courses		环境认知 Environmental Cognition	古建筑测绘 Survey and Drawing of Ancient Building	工地实习 Practice in Construction Plant

研究生培养（硕士学位）Graduate Program (Master Degree)			研究生培养（博士学位）
一年级 1st Year	二年级 2nd Year	三年级 3rd Year	Ph. D. Program

学术研究训练 Academic Research Training

学术研究 Academic Research

建筑设计研究 Research of Architectural Design	毕业设计或学位论文 Thesis Project or Dissertation	学位论文 Dissertation

专业核心理论 Core Theory of Architecture	专业扩展理论 Architectural Theory Extended	专业提升理论 Architectural Theory Upgraded	跨学科理论 Interdisciplinary Theory

建筑构造实验室 Building Construction Lab

建筑物理实验室 Building Physics Lab

数字建筑实验室 CAAD Lab

课程安排
CURRICULUM OUTLINE

	本科一年级 Undergraduate Program 1st Year	本科二年级 Undergraduate Program 2nd Year	本科三年级 Undergraduate Program 3rd Year
设计课程 Design Courses	设计基础 Design Foundation	建筑设计基础 Architectural Design Foundation 建筑设计（一） Architectural Design 1 建筑设计（二） Architectural Design 2	建筑设计（三） Architectural Design 3 建筑设计（四） Architectural Design 4 建筑设计（五） Architectural Design 5 建筑设计（六） Architectural Design 6
专业理论 Architectural Theory		建筑导论 Introductory Guide to Architecture 城乡规划原理 Theory of Urban and Rural Planning	建筑设计基本原理 Basic Theory of Architectural Design 居住建筑设计与居住区规划原理 Theory of Housing Design and Residential Planning
建筑技术 Architectural Technology		Python程序设计 Python Programming 理论力学 Theoretical Mechanics	建筑技术（一） Architectural Technology 1 建筑技术（二）声光热 Architectural Technology 2 Sound, Light and Heat 建筑技术（三）水电暖 Architectural Technology 3 Water, Electricity and Heating
历史理论 History Theory		外国建筑史（古代） History of Western Architecture (Ancient) 中国建筑史（古代） History of Chinese Architecture (Ancient)	外国建筑史（当代） History of Western Architecture (Modern) 中国建筑史（近现代） History of Chinese Architecture (Modern)
实践课程 Practical Courses		古建筑测绘 Survey and Drawing of Ancient Building	工地实习 Practice in Construction Plant
通识类课程 General Courses	数学 Mathematics 英语 English 思想政治 Ideology and Politics 科学与艺术 Science and Art	社会学概论 Introduction of Sociology	
选修课程 Elective Courses	人居环境导论 Research Method of the Social Science 现代工程与应用科学导论 Introduction to Modern Engineering and Applied Science 系统、决策与控制导论 Introduction to Systems, Decision-Making and Control 普通物理（力学） General Physics (Mechanics) 大学化学 College Chemistry 管理学 Management 经济学原理 Principles of Economics 自动化导论 Introduction to Automation	CAAD理论与实践 Theory and Practice of CAAD 城市道路交通规划与设计 Planning of Urban Roads and Traffic 环境科学导论 Introduction to Environmental Science	

本科四年级	研究生一年级	研究生二、三年级
Undergraduate Program 4th Year	Graduate Program 1st Year	Graduate Program 2nd & 3rd Years
建筑设计（七） Architectural Design 7 建筑设计（八） Architectural Design 8 本科毕业设计 Graduation Project	建筑设计研究（一） Architectural Design Research 1 建筑设计研究（二） Architectural Design Research 2 研究生国际教学工作坊 Postgraduate International Design Studio	专业硕士毕业设计 Thesis Project
城市设计及其理论 Urban Design and Theory	城市形态与设计方法论 Urban Morphology and Design Methology 建筑与规划研究方法 Research Method of Architecture and Urban Planning 现代建筑设计基础理论 Preliminaries in Modern Architectural Design	
建筑师业务基础知识 Introduction of Architects' Profession 建设工程项目管理 Management of Construction Project	建筑体系整合 Building System Integration 建筑学中的技术人文主义 Technology of Humanism in Architecture 建筑环境学与设计 Architectural Enviromental Science and Design GIS基础与应用 Concepts and Application of GIS	
	建筑理论研究 Studies of Architectural Theory	
		建筑设计实践 Architectural Design and Practice
景观规划设计及其理论 Landscape Planning and Design and Theory 建筑节能与绿色建筑 Building Energy Efficiency and Green Building	景观都市主义理论与方法 Theory and Method of Landscape Urbanism 建筑史研究 Studies of the History of Architecture 材料与建造 Materials and Construction 中国木建构文化研究 Studies in Chinese Wooden Tectonic Culture 计算机辅助技术 Computer Aided Design 传热学与计算流体力学基础 Fundamentals of Heat Transfer and Computational Fluid Dynamics 建设工程项目管理 Management of Construction Project 数字建筑设计 Digital Architecture Design	

2
设计基础
DESIGN FOUNDATION

12
建筑设计基础
ARCHITECTURAL DESIGN FOUNDATION

16
建筑设计（一）：限定与尺度——独立居住空间设计
ARCHITECTURAL DESIGN 1—LIMITATION AND SCALE: INDEPENDENT LIVING SPACE DESIGN

20
建筑设计（二）：展览空间与感知——文怀恩旧居加建设计
ARCHITECTURAL DESIGN 2—EXHIBITION SPACE AND PERCEPTION: EXTENSION DESIGN OF WEN HUAI'EN'S FORMER RESIDENCE

24
建筑设计（三）：专家公寓设计
ARCHITECTURAL DESIGN 3: THE EXPERT APARTMENT DESIGN

28
建筑设计（四）：世界文学客厅
ARCHITECTURAL DESIGN 4: WORLD LITERATURE LIVING ROOM

34
建筑设计（五）：大学生健身中心改扩建设计
ARCHITECTURAL DESIGN 5: RECONSTRUCTION AND EXPANSION DESIGN OF COLLEGE STUDENT FITNESS CENTER

42
建筑设计（六）：社区文化艺术中心设计
ARCHITECTURAL DESIGN 6: DESIGN OF COMMUNITY CULTURE AND ART CENTER

50
建筑设计（七）：高层办公楼设计
ARCHITECTURAL DESIGN 7: DESIGN OF HIGH-RISE OFFICE BUILDINGS

56
建筑设计（八）：城市设计
ARCHITECTURAL DESIGN 8: URBAN DESIGN

62
本科毕业设计
GRADUATION PROJECT

目 录

72
建筑设计研究（一）：基本设计
ARCHITECTURAL DESIGN RESEARCH 1: BASIC DESIGN

84
建筑设计研究（一）：概念设计
ARCHITECTURAL DESIGN RESEARCH 1: CONCEPTUAL DESIGN

96
建筑设计研究（二）：综合设计
ARCHITECTURAL DESIGN RESEARCH 2 : COMPREHENSIVE DESIGN

102
建筑设计研究（二）：城市设计
ARCHITECTURAL DESIGN RESEARCH 2 : URBAN DESIGN

114
研究生国际教学工作坊
POSTGRADUATE INTERNATIONAL DESIGN STUDIO

1—129 课程概览 COURSE OVERVIEW

131—143 建筑设计课程 ARCHITECTURAL DESIGN COURSES

145—147 建筑理论课程 ARCHITECTURAL THEORY COURSES

149—151 城市理论课程 URBAN THEORY COURSES

153—155 历史理论课程 HISTORY THEORY COURSES

157—160 建筑技术课程 ARCHITECTURAL TECHNOLOGY COURSES

161—167 其他 MISCELLANEA

课程概览
COURSE OVERVIEW

设 计 基 础
DESIGN FOUNDATION

鲁安东 唐莲 梁宇舒 尹航
LU Andong, TANG Lian, LIANG Yushu, YIN Hang

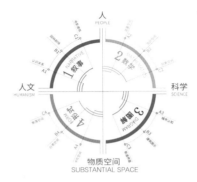

教案设计的背景
当代大学教育趋向将博学与精专相统一的通识教育，建筑学的职业化教学体系需要同时满足通识教育的现代多元化育人要求。针对这一目标，本教案提出了新的设计基础教学体系，为创意工科大类提供核心的思维训练和能力培养。本课程为工科大类通识课，选课人数为80—100人，其中约三分之一的学生后续选择建筑学方向。

培养多元融合的"设计思维"
本教案将创意工科中的"设计基础"理解为在人与物质空间之间、综合人文与科学的多元途径的创造性实践，因此教学体系确定以下4条主题线索：
主题一 叙事：关注物质空间中个体人的感受和意义。
主题二 数学：关注物质空间中人的共性和规律。
主题三 形式：关注物质空间对人的容纳与支持。
主题四 图解：关注物质空间本身及其运行状态。
这四条主题线索各有侧重，帮助学生全面理解"设计"的本质，并为学生建立"设计思维"提供多元融合的整体视野和框架。

能力导向的培养路径
通识教育注重能力培养而不是技能培养。本教案将创意工科设计基础的核心能力分解为3个部分并引导学生逐步加以学习：
阶段一 感受认知能力：对个体人或者物质空间进行多角度、多形式的观察、记录、描述，关注对问题建立整体和理性的认识。
阶段二 分析转化能力：在感受认知的基础上，在人与物质空间、人文与科学的整体视野和框架下，关注对具体问题的分析与转化。
阶段三 创造设计能力：在认知与转化的前提下，创造性地提出、分析和处理问题，关注从思维到行动的实施与完成。

基于自主学习的模块化教学
本教案基于4条主题线索和3个能力培养阶段，设计了12个时长5周的教学模块。学生可以根据自己的兴趣和需求自由选修不同模块，量身塑造自己的设计思维和设计能力。通过模块化教学，本教案发挥了通识教育下自主学习的优势，开展理性、全面的思维训练，突出系统、多元的能力培养。通过将设计基础作为创意工科的"元"学科，既为学生进一步的专业学习打下扎实基础，也培养了学生未来跨学科创新的必要素质。

Design background of the teaching plan
Contemporary university education tends to adopt liberal education integrating extensive and specialized knowledge. The professional teaching system of architecture should also meet the requirements for modern diversified education of liberal education. In response to this objective, this teaching plan proposes a new design foundation teaching system, which can provide core thinking training and competence training for creative engineering. This course, as a general course in engineering, may admit 80-100 students, about one-third of whom would select architecture in the future.

Cultivation of multivariate "design thinking"
In this teaching plan, the "design foundation" in creative engineering is interpreted as creative practice integrating humanities and science between human and substantial space; therefore, this teaching system centers on four thematic clues:
Theme 1 Narration: Pay attention to individual feeling and significance in substantial space.
Theme 2 Mathematics: Pay attention to people's commonality and laws in substantial space.
Theme 3 Form: Pay attention to accommodation and support for people in substantial space.
Theme 4 Illustration: Pay attention to the substantial space itself and its operating state.
These four thematic clues focus on different aspects, with the aim of helping the students to fully understand the essence of "design", and providing a multivariate overall vision and framework for the establishment of "design thinking" among the students.

Competence-oriented training path
Liberal education focuses on competence training rather than skill cultivation. In this teaching plan, the core competence of design foundation of creative engineering is divided into three parts, and students are guided to engage in gradual learning:
Stage 1 Perception and cognition competence: Observe, record, and describe individual people or substantial space in multiple angles and forms, and focus on establishing a holistic and rational understanding of problems.
Stage 2 Analysis and transformation competence: On the basis of perception and cognition, and under the overall vision and framework of human and substantial space, as well as humanity and science, focus on the analysis and transformation of specific issues.
Stage 3 Creative design competence: Under the premise of cognition and transformation, creatively propose, analyze and deal with problems, and focus on implementing and completing the process from thinking to action.

Modular teaching based on autonomous learning
In this teaching plan, 12 five-week teaching modules are designed based on the four thematic clues and three competence training stages. Students can freely select different modules according to their own interest and needs, to tailor their design thinking and design competence. Through modular teaching, this teaching plan can make use of the advantages of autonomous learning under liberal education, thus carrying out rational and comprehensive thinking training, and highlighting systematic and multivariate competence training. Design foundation, the "meta" subject of creative engineering, can lay a solid foundation for professional learning for students, and cultivate the necessary competences for interdisciplinary innovation in the future.

感受认知	分析转化	创造设计
5周（个人作业） 每组20人左右	5周（个人作业） 每组20人左右	5周（个人作业） 每组20人左右
A1 鲁安东 园林剧场 	B1 鲁安东 影像建筑 	C1 鲁安东 南大校园场所振兴计划
A2 唐莲 水泥容器 	B2 唐莲 桥 	C2 唐莲 设计基础 
A3 梁宇舒 住屋文化 	B3 梁宇舒 逻辑生长 	C3 梁宇舒 聚落重构
A4 尹航 街道界面认知 	B4 尹航 建筑空间认知 	C4 尹航 城市认知图示

设计基础 DESIGN FOUNDATION
B1：影像建筑
B1:BUILDINGS FOR IMAGE

鲁安东
LU Andong

教学进程
第一周
    模块简介
    讲课"日常性的影像博物馆"
    课后作业：自选三个电影片段，寻找电影中的一个日常建筑元素：楼梯、走廊、窗、门、阳台、露台、庭院，并配文描述建筑元素的空间特质。

第二周
    上周作业讲评
    课后任务解析
    课后作业：对建筑元素做出定义，运用建筑元素的重复，塑造一个无尽的空间。

第三周
    上周作业讲评
    课后任务解析
    课后作业：基于无尽的空间，叠加一种新的功能，运用建筑元素，将这个空间转变为建筑。

第四周
    上周作业讲评
    课后任务解析
    课后作业：优化建筑形式，完善细节。

Teaching process
Week 1
Module introduction
Lecture "Daily Image Museum"
Homework: Select three movie clips, and look for daily architectural elements, such as stairs, corridors, windows, doors, balconies, terraces, and courtyards, and describe their spatial characteristics.

Week 2
Comment on homework of the last week
After-class task analysis
Homework: Define the architectural elements, and create the endless space with repeated architectural elements.

Week 3
Comment on homework of the last week
After-class task analysis
Homework: Based on the endless space, superimpose a new function and transform the space into a building with the architectural elements.

Week 4
Comment on homework of the last week
After-class task analysis
Homework: Optimize the architectural form and improve the details.

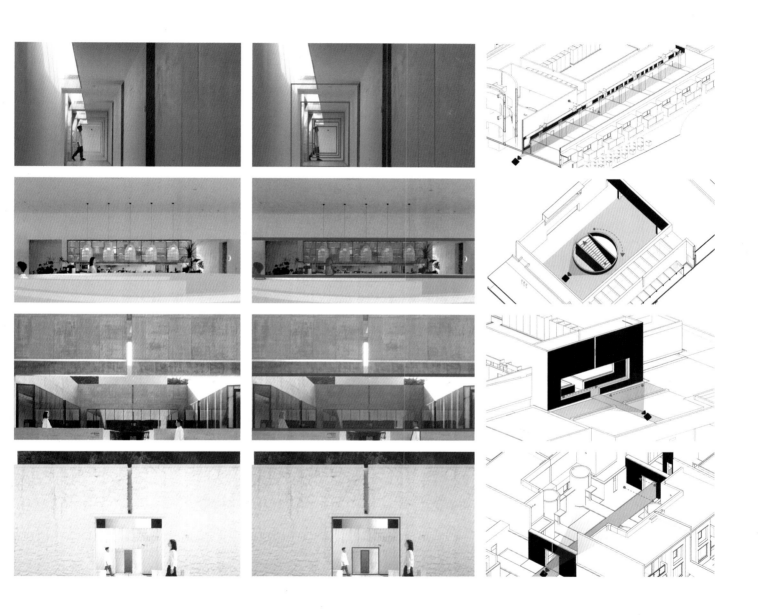

学生：陈琤 何德林 索朗顿珠　Student: CHEN Cheng, HE Delin, Sorangdunzhu

设计基础 DESIGN FOUNDATION

C1：南大校园场所振兴计划
C1: NANJING UNIVERSITY CAMPUS SPACE-MAKING

鲁安东
LU Andong

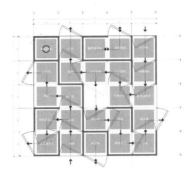

课程内容

本模块针对仙林校区教学楼内的消极空间进行研究，并提出场所提升方案，使其更加人性化，同时探讨信息技术和学习方式的改变对校园生活和交往空间的影响。本作业包括3个阶段。

阶段一：研究观察方法（2周）

学生3人一组，对教学楼消极空间的使用和行为进行2h观察，并对行为、环境感知、事件进行记录。4组学生分别以今和次郎的考现学、威廉姆·H.怀特的《小城市空间的社会生活》、萨拉·威格尔斯沃斯和杰里米·蒂尔的《餐桌越趋混乱》、伯纳德·屈米的《曼哈顿手稿》为方法，观察同一个场地，并对观察结果进行分析和制图。

阶段二：设计搭建方案（2周）

第一周学生3人一组，提出搭建方案，进行测试论证，并提出实施计划。第二周4组学生分别从形态、材料、行为、交互四个角度对最终方案提出要求，并在最终方案中进行整合。

阶段三：现场搭建装置（1周）

学生集体完成实地搭建，并记录场所营造效果。
搭建场地：南京大学仙林校区仙2教学楼2F公共空间。

Course content

This module studies the negative space in the teaching building of Xianlin campus, and proposes a place improvement plan to make it more hommization. At the same time, it discusses the impact of information technology and changes in learning methods on campus life and communication space. This assignment consists of 3 stages.

Stage 1: Research and observation methods (2 weeks)

A group of 3 students will observe the use and behavior of the negative space in the teaching building for 2 hours and record their behavior, environment perception, and event. The 4 groups of students use a method based on Imawa Jiro's Modernology, William H. Whyte's *The Social Life of Small Urban Spaces*, Sarah Wigglesworth and Jeremy Till's *Increasing Disorder in a Dining Table*, and Bernard Chimi's *Manhattan Manuscript*, observe the same site, and analyze and graph the observation results.

Stage 2: Design and build plan (2 weeks)

In the first week, a group of 3 students proposes a construction plan, conduct a test and demonstration, and proposes an implementation plan. In the second week, the 4 groups of students put forward requirements on the final plan from the four perspectives of forms, materials, behaviors, and interaction, and integrate them in the final plan.

Stage 3: On-site installation (1 week)

Students collectively complete the on-site construction and record the effect of the place construction.
Construction site: Public space on 2F, Xian 2 Teaching Building, Xianlin Campus, Nanjing University.

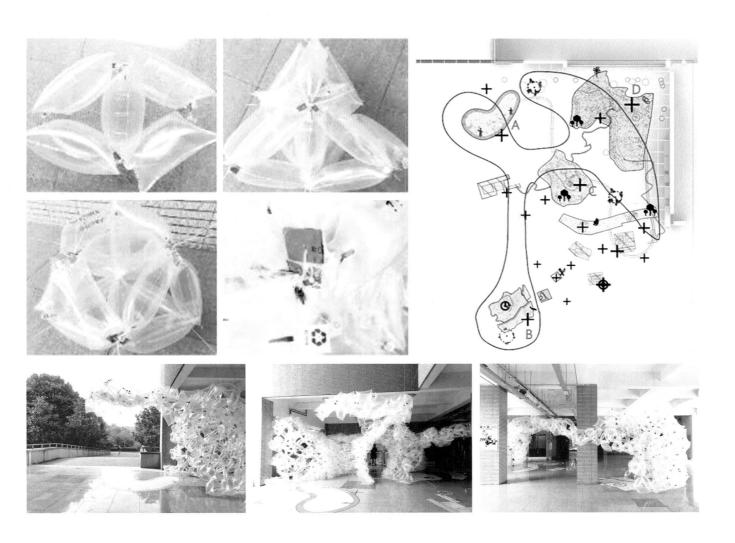

学生：许妍 龚晨宇 张嘉木 熊婧怡 陈沈婷 黄译文 陈琤 何德林 王玉萌 刘珩歆 沈至文 柴鑫
Student: XU Yan, GONG Chenyu, ZHANG Jiamu, XIONG Jingyi, CHEN Shenting, HUANG Yiwen, CHEN Cheng, HE Delin, WANG Yumeng, LIU Hengxin, SHEN Zhiwen, CHAI Xin

设计基础 DESIGN FOUNDATION

B3：逻辑生长
B3: LOGICAL GROWTH

梁宇舒
LIANG Yushu

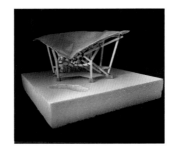

教学目标
通过对模块 A 所研究的人类原始住屋案例中的一类住屋进行认知，培养学生从已有的建筑材料语言、结构语言和气候语言中提取单一项，进行以简单公共装置设计为目标的立体构成创作，从而达到从要素提取、原理分析到形式演绎的设计训练（对乡土建筑进行材料使用、结构技艺以及气候原理层面的挖掘）。

教学内容
"逻辑生长"模块通过向同学介绍乡土住宅的材料、建造、气候适应性等特点，引导学生在案例研究的基础上，关注建构逻辑，进行立体构成学层面的实践，包含逻辑提取与设计再演绎的手工模型设计与制作。

教学过程历时 5 周（含评图 1 周），包括 3 个练习：

A：2 人一组，基于模块 A 的研究对象，进行对案例对象的结构、构件、材料及气候适应性层面的文献查找；

B：2 人一组，基于以上研究，抽取单一要素（结构／构造／适应性），进行相应的原理分析（图示分析或手模分析）；

C：2 人一组，利用立体构成学原理，设计出发点为一个可以置于城市广场的装置或提供公共活动的雨篷，完成手工模型（25 cm×25 cm×25 cm）的制作。从而进行一场小型"地域建筑实践"。

教学进程
第一周：1）讲课——乡土建筑之材料、结构与建造；2）布置作业 A。

第二周：1）汇报讲评作业 A（1 h）；2）讲课——材料的使用及模型的制作；3）布置作业 B。

第三周：1）汇报讲评作业 B（1 h）；2）讲课——艺术·设计的立体构成；3）布置作业 C。

第四周：1）汇报模型制作方案（1 h）；2）指导模型制作。

第五周：评图。

Teaching objectives
Through a class of houses in the human original housing case studied in Module A, cultivate students to extract individuals from existing building materials, structural language, and climate language. Carry out stereoscopic composition creation aiming at the design of simple public devices, thereby achieving design training from elements extraction, principle analysis to formal deduction (Excavate the material use, structural skills, and climate principle levels of native building).

Teaching content
The "Logical Growth" module introduces students to the materials, construction, climate adaptability, and other characteristics of rural houses, and guides students to pay attention to the construction logic on the basis of case studies and to carry out the three-dimensional composition level, including the manual extraction of logic and the reinterpretation of design on model design and production.

The teaching process lasts five weeks (including one week of commenting on pictures), including three exercises:

A: A team of two, based on the research object of module A, search for the literature on the structure, component, material and climate adaptability of the case object;

B: A team of two, based on the above research, extract a single element (structure/construction/adaptability), and carry out the corresponding principle analysis (graphic analysis or hand mold analysis);

C: A group of two people, using the principle of three-dimensional composition, the starting point of the design is a device that can be placed in a city square or an awning for public activities, and the production of handmade models (25 cm×25 cm×25 cm) is completed. Thus, a small "regional architecture practice" is carried out.

Teaching process
Week 1: 1) Lecture—material, structure and construction of vernacular architecture; 2) assign homework A.

Week 2: 1) Report and comment on homework A (1 h); 2) lecture—the use of materials and the making of models; 3) assign homework B.

Week 3: 1) Report and comment on homework B (1 h); 2) lecture—stereoscopic composition of art and design; 3) assign homework C.

Week 4: 1) Report the model making plan (1); 2) instruct model making.

Week 5: Commenting on pictures.

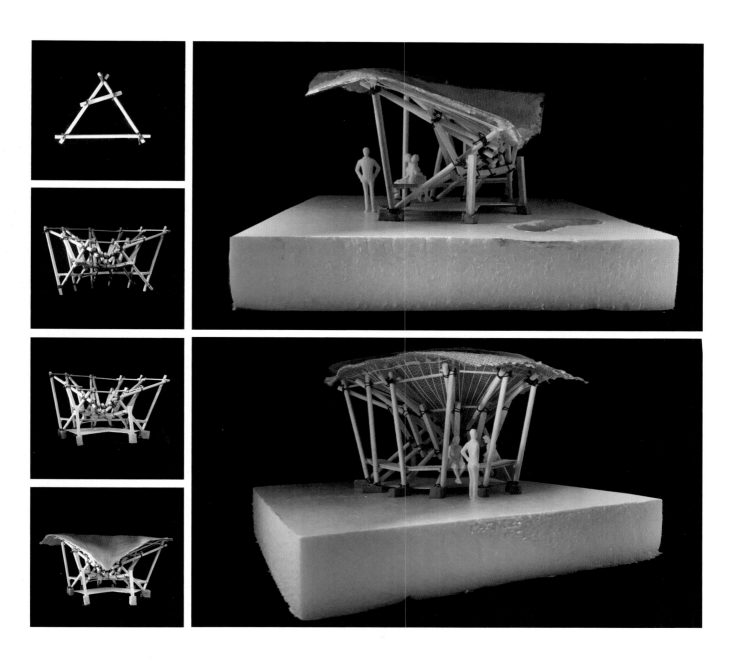

学生：沈至文　Student: SHEN Zhiwen

设计基础 DESIGN FOUNDATION

C3：聚落重构
C3: COMMUNITY SPACE RECONSTRUCTION

梁宇舒
LIANG Yushu

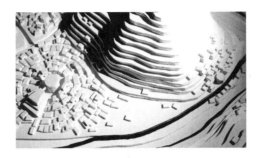

教学目标

培养从基本空间单元到单元组合的场所设计和社区构建的设计能力（基于地域要素的新型社区的逻辑重构）。

教学内容

"社区重构"模块培养学生关注住屋与住屋之间的空间关系处理、路径处理、结构组织处理，鼓励再现原有聚落的文化信仰、地理特征、场所意向等多要素中的一个或多个，在给定框架内设计一座"新型社区"。

教学过程历时5周（含评图1周），包括3个练习：

A：2人一组，以前一阶段的单元设计为基础，制订建筑单元体的标准模块选形、块数计划，要有1—2个社区中心。

B：2人一组，制作1：1000场地模型，根据地域要素地理特征保留一类场地要素。

C：2人一组，针对一个典型聚落空间元素，抽取单一要素，在标准模块上于空置的场地上设计一个小型社区环境。

教学进程

第一周：1）讲课——有趣的聚落形态、当代社区的重构；2）布置作业A。
第二周：1）汇报讲评作业A（1h）；2）布置作业B。
第三周：1）汇报讲评作业B（1h）；2）布置作业C。
第四周：汇报讲评作业C（1h）。
第五周：评图。

Teaching objectives

Cultivate the design ability of site and community design from basic space unit to unit combination (Logical reconstruction of new community based on geographical elements).

Teaching content

Community reconstruction module is mainly used to cultivate the students to focus on the handling of spatial relationship between houses, the processing of path and the treatment of structural organization; and encourage them to reproduce one or several factors such as the cultural beliefs, geographic features, and site intentions of the original settlement, so as to design a new community within a given framework.

The teaching process lasts for five weeks (including drawing review for one week) and consists of three exercises:

A: Two students in a group, based on the unit design of the previous stage, formulate the standard module selection and block number plan for the building unit body, and there must be 1-2 community centers.

B: Two students in a group, make a site model, 1：1000, and reserve a class of site elements according to the geographical features of regional elements.

C: Two students in a group, for a typical settlement space element, extract the single element, and design the small community environment on a vacant site on a standard module.

Teaching process

Week 1: 1) Lectures—interesting settlement forms and reconstruction of contemporary communities; 2) assign homework A.
Week 2: 1) Report and comment on homework A (1 h); 2) assign homework B.
Week 3: 1) Report and comment on homework B (1 h); 2) assign homework C.
Week 4: Report and comment on homework C (1 h).
Week 5: Drawing evaluation.

学生：吴琦　Student: WU Qi

建筑设计基础
ARCHITECTURAL DESIGN FOUNDATION

刘铨　史文娟
LIU Quan, SHI Wenjuan

教案设计的背景

"建筑设计基础"课程的主要任务是在中国建筑教育的现实条件下，让原本对建筑学一无所知的新生建立基础性的专业知识架构。其主要内容是建筑认知和建筑表达。认知是主线，表达是方法。认知成果须通过表达方式得以检验，而表达的效果和认知成果直接对应。以这一目标为出发点，本教案提出了建筑设计的基础教学体系，将基础教学分解为针对建筑对象的三个基本建筑问题，即空间与尺度、结构与构造、场地与环境。

本课程为专业核心课程，教学对象为本科二年级，课程人数为30人。

教案架构

教案的基本架构是在重新认识建筑基础知识的前提下，将认知与表达作为这门课的教学主线，依照循序渐进的原则，分三个阶段设置了不同的教学任务，每个阶段有其特定的认知对象和认知方法，包含若干练习。同时每个阶段的训练都建立在之前一个阶段学习要点的基础上，力图更好地使学生通过认知的过程从一个外行逐步进入专业领域，并为后续的建筑设计学习打下宽阔和扎实的基础。

阶段一：
建筑立面局部测绘：从材料与构件尺寸认知到正投影图绘制；
建筑物测绘：从建筑空间分割与功能尺度认知到平、剖面图绘制；
建筑窗测绘：从建筑构造认知到大样图绘制。
阶段二：
建筑结构模型制作：从结构图识图到结构模型制作；
墙身模型制作：从大样图识图到构造模型制作。
阶段三：
街道空间认知：理解街巷肌理、城市街道空间及其限定与功能；
地块与建筑类型认知：理解地块肌理、城市建筑类型及其功能与交通组织；
地形与气候认知：理解自然地形、植被及日照等自然环境要素。

教学方法

线上线下结合：本课程使用了慕课作为线上教学工具，方便学生对知识的学习，课堂则解决应用中的具体问题和进行实操指导，同时尽量利用网络分发资料和提交作业。

课堂课外结合：测绘、结构构造认知、城市空间认知阶段都设置了现场讲解，使学生直接面对学习对象，提高了教学效果。

Background of teaching plan design

The main task of the course "Architectural Design Foundation" is to establish a basic professional knowledge framework for freshmen who originally know nothing about architecture under the realistic conditions of Chinese architectural education. Its main content is architectural cognition and architectural expression. Cognition is the main line and expression is the method. Cognitive achievements need to be tested by means of expression, and the effect of expression corresponds directly to cognitive achievements. Taking this goal as the starting point, this teaching plan puts forward the basic teaching system of architectural design, which divides the basic teaching into three basic architectural problems for architectural objects, namely space and scale, structure and construction, site and environment.

This course is a professional core course, the teaching object is the second year of undergraduate, and the student number of courses is 30.

Teaching plan structure

The basic structure of the teaching plan is to take cognition and expression as the main teaching line of this course under the premise of reunderstanding the basic knowledge of architecture. According to the principle of step-by-step, different teaching tasks are set in three stages. Each stage has its specific cognitive objects and cognitive methods, including several exercises. At the same time, the training of each stage is based on the learning points of the previous stage, trying to better enable students to gradually enter the professional field from a layman through the cognitive process, and lay a broad and solid foundation for subsequent architectural design learning.

Stage 1:
Local surveying and mapping of building facade: From recognizing the size of materials and components to drawing orthographic projection.
Building surveying and mapping: Recognizing the plan and section drawing from building space segmentation and functional scale.
Building window surveying and mapping: From building structure cognition to detail drawing.

Stage 2:
Building structural model making: From structural drawing recognition to structural model making.
Wall body model making: from detail drawing identification to structural model making.

Stage 3:
Street space cognition: Understanding street texture, urban street space and its limitations and functions.
Cognition of the plot and building type: Understanding plot texture, urban building type, and its function and traffic organization.
Terrain and climate cognition: Understanding natural environment elements such as natural terrain, vegetation and sunshine.

Teaching methods

Combination of online and offline: This course uses Mooc as an online teaching tool to facilitate students' learning of knowledge. In class, it solves specific application questions and gives practical guidance. At the same time, it also makes full use of the network to solve data distribution and homework submission.

Classroom and extracurricular combination: On-site explanation is set for all three phases of surveying and mapping, structural cognition and urban space cognition. Students face the learning objects directly, which improves the teaching effect.

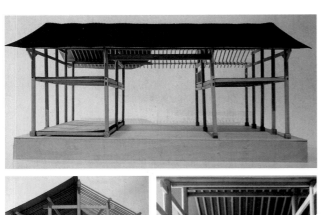

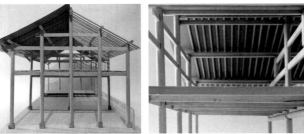

学生：夏月 Student: XIA Yue

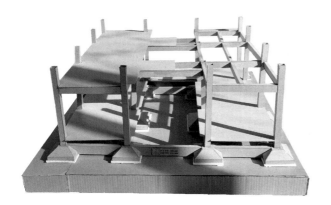
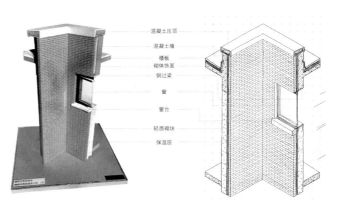

学生：顾靓 Student: GU Liang

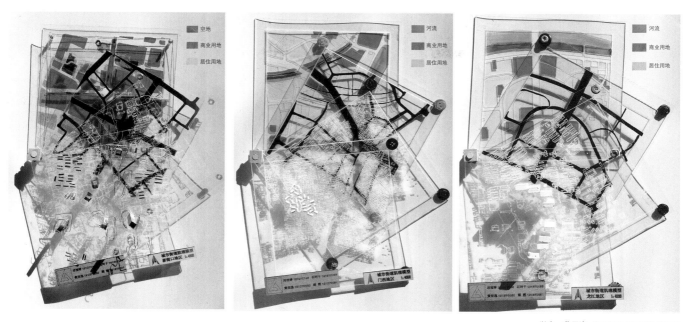

学生：黄辰逸　Student: HUANG Chenyi

学生：石珂千　Student: SHI Keqian

建筑设计（一） ARCHITECTURAL DESIGN 1

限定与尺度——独立居住空间设计
LIMITATION AND SCALE—INDEPENDENT LIVING SPACE DESIGN

刘铨　冷天　吴佳维
LIU Quan, LENG Tian, WU Jiawei

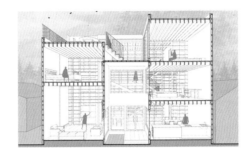

教学目标

本次练习的主要任务是综合运用前期案例学习中的知识点——建筑在水平方向上如何利用高度、开洞等操作划分空间，内部空间的功能流线组织及视线关系，墙身、节点、包裹体系、框架结构的构造方式，周围环境对空间、功能、包裹体系的影响等，初步体验一个小型独立居住空间的设计过程。

基本任务

1) 场地与界面：本次设计的场地面积为 80—100 m²，场地单面或相邻两面临街，周边为 1—2 层的传统民居。

2) 功能与空间：本次设计的建筑功能为小型家庭独立式住宅（附设有书房功能）。家庭主要成员包括一对年轻夫妇和 1—2 位儿童（7 岁左右）。新建建筑面积 160—200 m²，建筑高度 ≤ 9 m（不设地下空间）。设计者根据设定的家庭成员的职业及兴趣爱好确定空间的功能（职业可以是但不局限于理、工、医、法的技术人员）。

3) 流线组织与出入口设置：考虑建筑内部流线合理性以及建筑出入口与场地周边环境条件的合理衔接。

4) 尺度与感知：建筑中的各功能空间的尺寸需要以人体尺度及人的行为方式作为基本的参照，并通过图示表达空间构成要素与人的空间体验之间的关系。

教学进度

本次设计课程共 6 周。

第一周：构思并撰写几个有代表性的生活场景（家庭人物构成、人物相对关系、类似戏剧中的"折"之剧本）。四个地块分别制作 1:100（60 cm×60 cm）的场地模型。利用 1:100 手绘图纸及 1:100 体块模型构思内部功能空间及其关系。

第二周：用 1:50 手绘平、立、剖面图纸，结合 1:50 工作模型辅助设计，在初步方案的基础上考虑功能与空间、流线与尺度。

第三周：确定设计方案，制作体现功能关系的空间关系模型（例如立体泡泡图），推进剖、立面设计。

第四周：设计深化，细化推敲各设计细节，并建模研究内部空间效果（集中挂图点评）。

第五周：制作 1:20 或 1:30 剖透视图和各分析图，制作 1:50 表现模型。

第六周：整理图纸、排版并完成课程答辩。

成果要求

A1 灰度图纸 2 张，纸质表现模型 1 个（1:50），场地模型 1 个（1:100），工作模型若干。图纸内容应包括：

1) 总平面图（1:200），各层平面图、纵横剖面图和主要立面图（1:50），墙身大样（1:10），内部空间组织剖透视图 1 张（1:20）。

2) 设计说明和主要技术经济指标（用地面积、建筑面积、容积率、建筑密度）。

3) 表达设计意图和设计过程的分析图（体块生成、功能分析、流线分析、结构体系等）。

4) 纸质模型照片与电脑效果图、照片拼贴等。

Teaching objectives

The main task of this exercise is to comprehensively use the knowledge points in the early case study—how to use height, opening and other operations to divide space in the horizontal direction of the building, functional streamline organization and line of sight relationship of internal space, construction mode of wall body, nodes, wrapping system and frame structure, the influence of the surrounding environment on the space, function and wrapping system, and preliminarily experience the design process of a small independent living space.

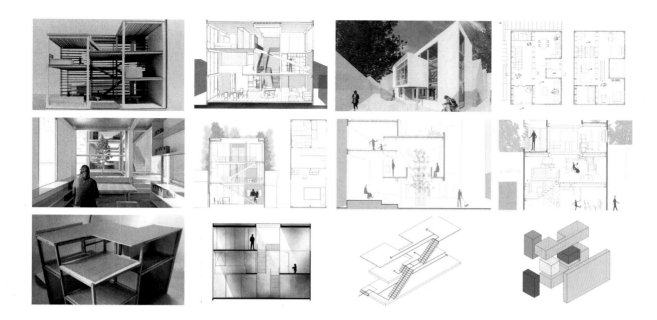

Basic tasks

1) Site and interface: The site of this design covers an area of about 80–100 m², facing the street on one side or two adjacent sides, surrounded by 1–2 floors of traditional residential buildings.

2) Function and space: The building function of this design is a small family independent residence (with study function attached). The main members of the family include a young couple and 1–2 children (about 7 years old). The new building area is 160–200 m² and the building height is ≤ 9 m (no underground space). The designer determines the function of the space according to the set occupation and interests of family members (the occupation can be but not limited to technicians of science, engineering, medicine and law).

3) Streamline organization and entrance and exit setting: Consider the rationality of the internal streamline of the building and the reasonable connection between the entrance and exit of the building and the surrounding environmental conditions of the site.

4) Scale and perception: The size of each functional space in the building needs to take the human body scale and human behaviors as the basic reference, and express the relationship between spatial constituent elements and human spatial experience through diagrams.

Teaching progress

This design course is 6 weeks in total:

Week 1: Conceive and write several representative life scenes (composition of family characters, relative relationship of characters, a script similar to "folding" in drama). Make 1∶100 (60 cm × 60 cm) site models for the four plots respectively. Use 1∶100 hand-painted drawings and the 1∶100 block model to construct the internal functional space and its relationship.

Week 2: 1∶50 hand drawn plan, elevation and section drawings, combined with 1∶50 working model aided design, considering function and space, streamline and scale on the basis of preliminary scheme.

Week 3: Determine the design scheme, make the spatial relationship model reflecting the functional relationship (such as three-dimensional bubble diagram), and promote the section and elevation design.

Week 4: Deepen the design, refine the design details, and model and study the internal space effect (centralized wall chart comments).

Week 5: Make (1∶20 or 1∶30) sectional perspective views and analysis diagrams, and make 1∶50 performance model.

Week 6: Organize drawings, typesetting and complete course defense.

Achievement requirements

Two A1 grayscale drawings, one paper performance model (1∶50), one site model (1∶100) and several working models. The contents of the drawings shall include:

1) General layout (1 : 200), plan of each floor, vertical and horizontal section and main elevation (1 : 50), wall detail (1 : 10), sectional perspective view of internal space organization (1∶20).

2) Design description and main technical and economic indicators (land area, building area, plot ratio and building density).

3) Analysis diagram expressing design intention and design process (block generation, function analysis, streamline analysis, structural system, etc.).

4) Paper model photos, computer renderings, photo collages, etc..

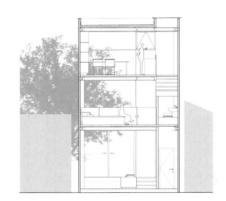
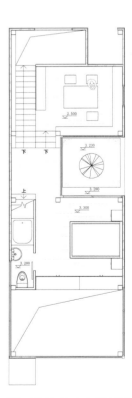
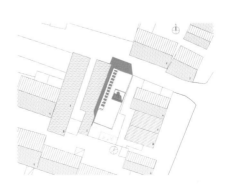

学生：黄辰逸　Student: HUANG Chenyi

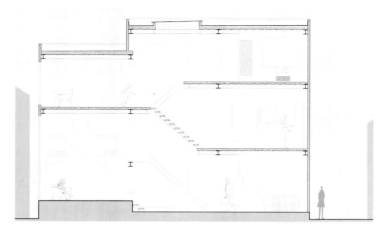
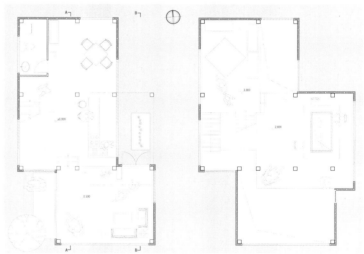

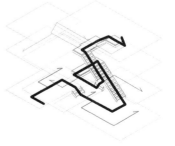

学生：邱雨婷　Student: QIU Yuting

建筑设计（二）ARCHITECTURAL DESIGN 2

展览空间与感知——文怀恩旧居加建设计
EXHIBITION SPACE AND PERCEPTION—EXTENSION DESIGN OF WEN HUAI'EN'S FORMER RESIDENCE

刘铨　冷天　吴佳维
LIU Quan, LENG Tian, WU Jiawei

教学目标
上一个设计题目关注的重点是利用水平构件（楼板）组织和限定建筑内部空间，利用楼板的大小、形状、开洞或错位关系形成所需的不同空间功能尺度与视觉联系。在本次设计中，我们需要进一步增加对场地环境的关注，注意新老建筑、内外空间、历史记忆与现实需求的关系，同时不仅利用水平构件，还要充分利用垂直构件（墙体、柱子）来组织和限定功能流线与营造视觉体验。

设计场地
设计训练的场地位于南京大学鼓楼校区新老轴线之间、教学楼东南侧文怀恩旧居所在的区域。文怀恩旧居的总建筑面积约为667 m²（包含地下室67 m²，阁楼73 m²），由主楼与附楼两部分组成，主入口朝东，面对金陵大学主轴线和小纪念公园。设计范围位于旧居东侧，面积大约为1 337.22 m²，包括了目前的一条南北向人行通路和公园的一部分，新建建筑红线则在设计范围的北侧，面积约为451.20 m²（详见地形与红线图）。

设计要求
1）功能要求：根据文怀恩与金陵大学的历史，设计一个展示纪念馆。老建筑由于缺少大空间，需要加设一个较大的灵活空间。建成后老建筑用作固定展陈和办公，新建筑则作为临时展览、研讨交流、会议茶歇等可以灵活使用的空间。同时由于其位于校园核心地带，新增建筑内拟设一个小型咖啡厅，为学校教职工及日常参观人群提供服务。新建建筑总面积不少于150 m²，建筑高度≤8 m（檐口高度，不包括女儿墙）。新建筑以一层为主，局部可设夹层。在场地内还应考虑一处与展览主题相关的纪念性空间。新老建筑应作为一个整体考虑其参观流线，但新建建筑也应考虑其相对独立性，在老建筑闭馆时可独立使用。

2）场地环境：现状建筑和场地内各项要素既是限制，又是形成新建筑体量的基本条件。本次设计场地南侧为小礼拜堂，北侧为教学楼，东侧面对小花园（原金陵大学主轴线上的入口花园）和金陵大学主轴线上的石碑、雕塑。结合与这些重要的环境要素的位置的视觉关系来考虑建筑物、纪念性空间的布局以及参观流线的组织。

3）空间限定要素与视觉关系的组织：本次训练需要通过空间限定要素（水平与垂直构件）与身体感知（路径、视线、活动与尺度、光影、质感）关系的研究，塑造出相应的室内外展陈与纪念性空间，连接文宅历史记忆与现实需求，创造性地再现该场所的人文内涵。

4）材料与建造：选择合适的材料、结构形式，呼应空间组织与场地环境需要。

教学进度
第一周：场地认知，结合已有图像资料、地形资料进行深度的场地认知。制作比例1∶50的场地模型。自拟文宅的校园新功能，写出详细的展陈和使用设定，包括活动场所服务对象的类别和数目、场地设施的具体要求，思考建筑策略的灵活性。通过案例分析，了解展陈空间的设计手法以及历史建筑复兴的改扩建操作思路。

第二周：提出处理文本设定相应的建筑策略，思考场地的新旧对话关系，整合空间的组织方式，形成初步方案与工作模型（1∶100）。

第三周：深化初步方案，优化并发展前述的策略，用1∶50的手绘平、立、剖面图纸，在初步方案的基础上深化功能与空间、流线与尺度（集中挂图点评）。

第四周：利用工作模型辅助设计，进一步优化设计，使得建筑结构清晰、明确、可认知，并研究内部空间效果，确定最终的设计方案。

第五周：深化1∶50图纸，细化推敲各设计细节，制作比例1∶50的模型（模型点评）。

第六周：方案优化，思考并选择图面表达的效果，制作必要的分析图和效果图。

第七周：整理图纸、排版，制作正式模型并完成课程答辩。

Teaching objectives
The last design topic focuses on the use of horizontal components (floors) to organize and limit the internal space of the building, and use the size, shape, opening or dislocation relationship of the floor to form the required different spatial functional scales and visual connections. In this design, we need to further pay attention to the site environment, pay attention to the relationship between new and old buildings, internal and external space, historical memory and practical needs, and make full use of not only horizontal components, but also vertical components (walls and columns) to organize and limit the functional streamline and create visual experience.

Design site
The design training site is located between the new and old axes of Gulou campus of Nanjing University and the former residence of Wen Huaien in the southeast of the teaching building. The total construction area of Wen Huai'en's former residence is about 667 m² (including 67 m² basement and 73 m² attic), which is composed of the main building and the auxiliary building. The main entrance faces

the east, facing the main axis of Jinling University and the small memorial park. The design scope is located in the east of the former residence, with an area of about 1 337.22 m^2, including a current north–south pedestrian access and part of the park. The new building red line is located in the north of the scope, with an area of about 451.20 m^2 (see topographic and red line map for details).

Design requirements

1) Functional requirements: Design an exhibition memorial hall according to the history of Wen Huai'en and Jinling University. Due to the lack of large space in the old building, a large flexible space needs to be added. After completion, the old building is used for fixed exhibition and offices, while the new building is used as a flexible space for the temporary exhibition, discussion and exchange, conference and tea break. At the same time, as it is located in the core of the campus, a small coffee shop is proposed to be set up in the new building to serve the school staff and daily visitors. The total area of new buildings shall not be less than 150 m^2, The building height is ⩽ 8 m (cornice height, excluding parapet). The newly-built building is mainly one floor, with the partial mezzanine. A memorial space related to the exhibition theme should also be considered in the site. The new and old buildings should be considered as a whole, but the new buildings should also consider their relative independence, which can be used independently when the old buildings are closed.

2) Site environment: The current buildings and various elements in the site are not only restrictions, but also the basic conditions for the formation of new building volume. The south side of the design site is the small chapel, the north side is the teaching building, and the east side faces the small garden (the entrance garden on the original Jinling University main axis) and the stone tablets and sculptures on the Jinling University main axis. Combined with the visual relationship with the location of these important environmental elements, the layout of buildings, memorial spaces and the organization of visiting streamlines are considered.

3) Organization of space limiting elements and visual relationship: This training needs to shape the corresponding indoor and outdoor exhibition and memorial space through the organization of the relationship between space limiting elements (horizontal and vertical components) and body perception (path, line of sight, activity and scale, light and shadow, texture), connect the historical memory and practical needs of the Wen House, and creatively reproduce the humanistic connotation of the place.

4) Materials and construction: Select appropriate materials and structural forms to meet the needs of space organization and site environment.

Teaching progress

Week 1: Site cognition, have the in-depth site cognition combined with existing image data and topographic data. Make a site model with a scale of 1∶50. Draw up the new campus functions of Wen House, write out the detailed exhibition and use settings, including the category and number of service objects in the activity place, the specific requirements of site facilities, and think about the flexibility of architectural strategies. Through case analysis, understand the design methods of exhibition space and the reconstruction and expansion operation ideas of the revival of historical buildings.

Week 2: Propose to deal with the text, set corresponding architectural strategies, think about the old and new dialogue relationship of the site, integrate the organization mode of space, and form a preliminary scheme and working model, with a ratio of 1∶100.

Week 3: Deepen the preliminary scheme, optimize and develop the above strategies, and use 1∶50 hand-drawn plan, elevation and section drawings to deepen the design on the basis of the preliminary scheme function and space, streamline and scale (centralized wall chart comments).

Week 4: Use the working model to assist the design, further optimize the design, make the building structure clear and recognizable, study the internal space effect and determine the final design scheme.

Week 5: Deepen the 1∶50 drawings, refine the design details, and make a 1∶50 model (model comments).

Week 6: Plan optimization, consider and select the effect of drawing expression, and make necessary analysis and effect drawings.

Week 7: Organize drawings, typesetting, make formal models and complete course defense.

学生：石珂千　Student: SHI Keqian

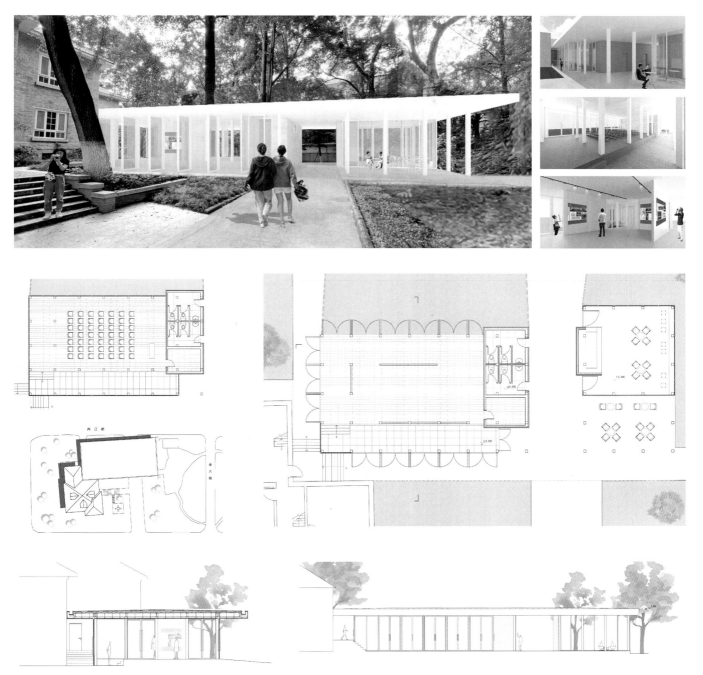

学生：黄辰逸　Student: HUANG Chenyi

建筑设计（三）ARCHITECTURAL DESIGN 3

专家公寓设计
THE EXPERT APARTMENT DESIGN

华晓宁　窦平平　黄华青
HUA Xiaoning, DOU Pingping, HUANG Huaqing

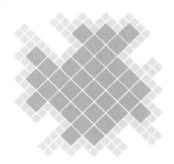

教学目标

从空间单元到系统的设计训练。

从个体到整体，从单元到体系，是建筑空间组织的一种基本和常用方式。本课题首先关注空间单元的生成，并进一步根据内在的使用逻辑和外在的场地条件，将多个单元通过特定方式与秩序组合起来，形成一个兼备合理性、清晰性和丰富性的整体系统。基本单元的重复、韵律、变异等都是常用的操作手法。

基本任务

拟在南京大学鼓楼校区南园宿舍区内新建专家公寓一座，用于国内外专家到访南大开展学术交流活动期间的居住。用地位于南园中心喷泉西侧，面积约3 600 m²。地块上原有建筑将被拆除，新建总建筑面积不超过3 000 m²。高度不超过3层。具体的功能空间包括：

客房：30间左右，分为单间和套间两类，单间面积为35—40 m²，套间面积为70—80 m²，内部需包括睡眠空间、卫生间、学习空间、工作空间。套间可考虑必要的接待空间和简单的餐厨空间。套间不少于5间。

大会议室1间：100—120 m²。

研讨室3间：每间约60 m²。

休闲区与咖啡吧（兼做早餐厅）：约150 m²。

操作间：约30 m²。

服务间：每层1间，每间约20 m²。

工作人员办公室与休息室：1—2间，每间约30 m²。

其他必要的门厅、前台、公共卫生间、储藏室、服务间等自行设置。

场地环境：结合建筑总体布局，在建筑周边及其内部创造优美的室内外场地环境，供使用者休憩交往，并为校园增色。

成果要求

图纸：总平面图（1：500），建筑平、立、剖面图（1：200），客房单元平面图（1：50），分析图、轴测图、鸟瞰图、剖透视图、人眼透视图，其他有助于表达方案的图纸。

手工模型：比例1：200。

教学进度

阶段一：场地调研分析，背景与案例研究，空间单元体的生成（2周）；

阶段二：单元的组合与总体布局（1周）；

阶段三：设计深化（3周）；

阶段四：最后成果制图、排版，准备答辩（2周）。

Teaching objectives

From space unit to system design training.

From individual to whole, from unit to system, is a basic and common way of architectural space organization. This topic first pays attention to the generation of spatial units, and further combines multiple units with order in a specific way according to the internal use logic and external site conditions to form an overall system with rationality, clarity and richness. Repetition, rhythm and variation of

basic units are commonly used.

Basic tasks

It is proposed to build a new expert apartment in the Nanyuan dormitory area of Gulou campus of Nanjing University for domestic and foreign experts to live during their visit to Nanjing University for academic exchange activities. The land is located in the west of the fountain in the center of Nanyuan, covering an area of about 3 600 m^2. The original buildings on the plot will be demolished, and the total construction area of the new buildings will not exceed 3 000 m^2. The height shall not exceed 3 floors. Specific functional space includes:

Guest rooms: about 30 rooms, divided into single rooms and suites, with a single room area of 35–40 m^2 and a suite area of 70–80 m^2. The interior needs to include sleeping space, toilet, study and work space. The necessary reception space and simple kitchen space can be considered in the suite. No less than 5 suites.
1 large conference room: 100–120 m^2.
3 seminar rooms: Each about 60 m^2.
Leisure area and coffee bar (also as breakfast restaurant): About 150 m^2.
Operation room: About 30 m^2
Service room: One room on each floor, about 20 m^2 each.
Staff offices and lounges: 1–2, each about 30 m^2.
Other necessary foyer, front desk, public toilet, storage room, service room, etc. shall be set by themselves.

Site environment: Combined with the overall layout of the building, create a beautiful indoor and outdoor site environment around and inside the building for users to rest and communicate, and enhance the landscape of the campus.

Achievement requirements

Drawings: General layout (1∶500), building plan, elevation and section (1∶200), guest room unit plan (1∶50), analysis drawing, axonometric drawing, aerial view, sectional perspective view, human eye perspective view, and other drawings which are helpful to express the scheme.
Manual model: Scale 1∶200.

Teaching progress

Stage 1: Site investigation and analysis, background and case study, generation of spatial unit (2 weeks);
Stage 2: Unit combination and overall layout (1 week);
Stage 3: Design deepening (3 weeks);
Stage 4: Final achievement drawing, typesetting and preparation of defense (2 weeks).

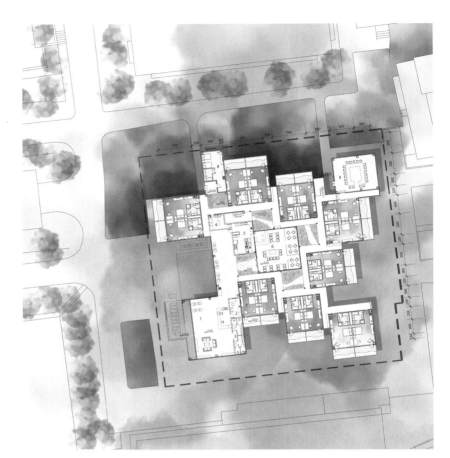

学生：田舒琳　Student: TIAN Shulin

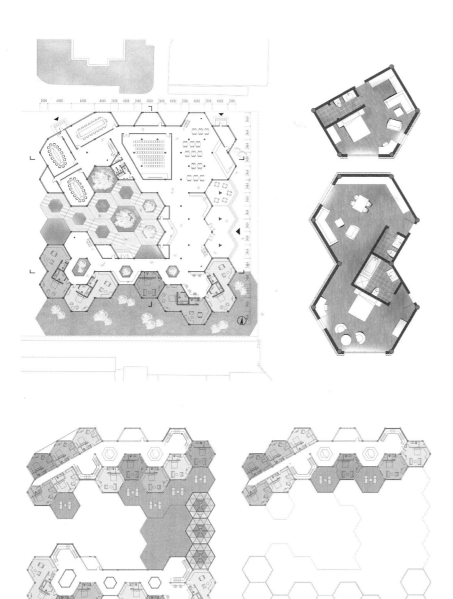

学生：李逸凡　Student: LI Yifan

世界文学客厅
WORLD LITERATURE LIVING ROOM

华晓宁　窦平平　黄华青
HUA Xiaoning, DOU Pingping, HUANG Huaqing

教学目标

本课程主题是"空间",学习建筑空间组织的技巧和方法,训练对空间的操作与表达。空间问题是建筑学的基本问题。本课题基于文学主题,训练文本、叙事与空间序列的串联,学习空间叙事与空间用途的整体构思,充分考虑人在空间中的行为、空间感受,尝试以空间为手段表达特定的意义和氛围,最终形成一个完整的设计。

设计场地

南京古称金陵、白下、建康、建邺……历来是人文荟萃、名家辈出之地,号称"天下文枢"。南京作为六朝古都,亦为中国文学之始。何为文？梁元帝曰:"吟咏风谣,流连哀思者,谓之文。"汉魏有文无学,六朝文学《文选》《文心雕龙》《诗品》既是文学评论的开始,也是文学的发端。

2019年,南京入选联合国"世界文学之都",开展一系列城市空间计划,包括筹建"世界文学客厅",作为一座以文学为主题的综合性博物馆。该馆选址位于北极阁公园东南隅,用地面积约5 050 m²,紧临市政府中轴线,毗邻古鸡鸣寺、玄武湖、明城墙、东南大学四牌楼校区等历史文化遗迹,构成城市与山林之间的过渡空间。设计应妥善处理建筑与周边城市环境和既有建筑的关系,彰显中国文学的精神特质。

空间计划

建筑总高度不超过12 m,容积率不超过0.5,绿地率不低于25%

空间组织要有明确特征和明确意图,概念清晰;满足功能合理、环境协调、流线便捷的要求。注意不同类型和不同形态空间的构成、空间的串联组织和空间氛围的塑造。

总建筑面积：约2 500 m²。

（以如下各部分面积配比为参考,每人可以根据研究自行策划并进行适当调整。）

1）世界文学之都展示中心
主题展厅：400—500 m²;
国际互联展厅：200 m²;
临时展厅：100—150 m²;
其他辅助空间（控制室、储藏间等）。

2）国际交流中心
报告厅（100座）：150—200 m²;
会议室（4—6间）：共300 m²;
会客厅：60—80 m²;
其他辅助空间（休闲、接待等）。

3）城市客厅
游客服务（可结合门厅）：150—200 m²。

4）旧书馆
书籍展示及阅览：150—200 m²。

5）行政办公与辅助空间
办公室：6间,每间不小于15 m²;
专家接待室：4间,每间不少于30 m²;
其他门厅、交通、设备间、卫生间等面积根据设计需要自行确定。

6）场地
园林景观、户外展场、停车场等,应考虑建筑与景观的整体关系,以景观烘托氛围。

Teaching objectives

The theme of this course is "space", learning the skills and methods of architectural space organization, and training the operation and expression of space. Space problem is the basic problem of architecture. Based on the literary theme, this topic trains the series of the text, narration and spatial sequence, learns the overall idea of spatial narration and spatial use, fully considers people's behaviors and spatial feelings in space, tries to express specific meaning and atmosphere by means of space, and finally forms a complete design.

Design site

In ancient times, Nanjing was called Jinling, Baixia, Jiankang and Jianye… It has always been a place with a large number of talents and famous scholars, known as the "cultural hub of the world". As the ancient capital for Six Dynasties, Nanjing is also the beginning of Chinese literature. What is literature? Emperor Liang Yuandi said, "it is called literature to chant wind ballads and linger around mourning." The Han and Wei Dynasties have literature but no learning. The literature of the Six Dynasties *Selected Works*, *Literary Heart and Carving Dragons* and *Poetry* are not only the beginning of literary criticism, but also the beginning of literature.

In 2019, Nanjing was selected as the "capital of world literature" of the United Nations and plans to carry out a series of urban space plans, including preparing to build the "world literature living room" as a comprehensive museum with literature as the theme. The museum is located in the southeast corner of Beijige Park, with a land area of about 5 050 m². It is close to the central axis of the municipal government and adjacent to historical and cultural sites such as ancient Jiming Temple, Xuanwu Lake, Ming City Wall and Sipailou campus of Southeast University, forming a transition space between the city and mountains. The design should properly deal with the relationship between the building and the surrounding urban environment and existing buildings, and highlight the spiritual characteristics of Chinese literature.

Space plan

The total height of the building shall not exceed 12 m, the plot ratio shall not exceed 0.5, and the green space rate shall not be less than 25%.

Spatial organization should have clear characteristics, clear intentions and clear concepts; meet the requirements of reasonable function, coordinated environment and convenient streamline. Pay attention to the composition of different types and forms of space, the series organization of space and the shaping of space atmosphere.

Total construction area: About 2 500 m².

(The area ratio of the following parts is for reference, and each person can make appropriate adjustment according to the research.)

1) Exhibition Center of the Capital of World Literature
Theme exhibition hall: 400–500 m²;
International internet exhibition hall: 200 m²;
Temporary exhibition hall: 100–150 m²;
Other auxiliary spaces (control room, storage room, etc.).

2) International exchange center
Lecture hall (100 seats): 150–200 m²;
Meeting rooms (4–6): 300 m² in total;
Reception rooms: 60–80 m² in total;
Other auxiliary spaces (leisure, reception, etc.).

3) City living room
Tourist service (can be combined with the lobby): 150–200 m².

4) Old library
Book display and reading: 150–200 m²;

5) Administrative office and auxiliary space
Office: 6 rooms, each not less than 15 m².
Expert reception room : 4 rooms, each not less than 30 m².

The area of other foyer, traffic, equipment room and toilet shall be determined according to the design needs.

6) Site
The overall relationship between architecture and landscapes shall be considered for the garden landscape, outdoor exhibition hall and parking lot, so as to set off the atmosphere with landscapes.

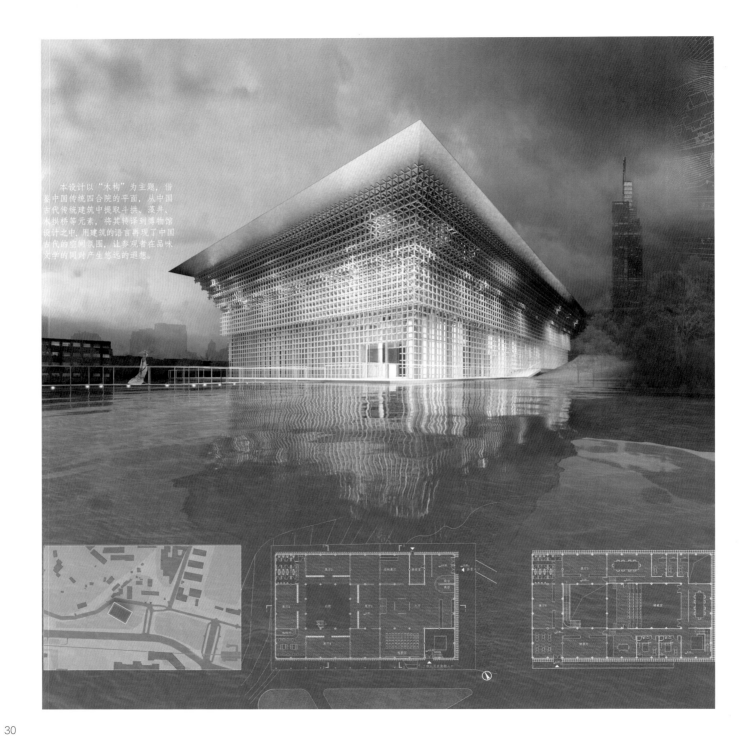

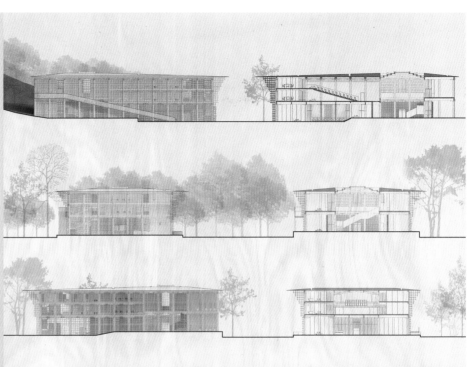

学生：李逸凡 Student: LI Yifan

学生：林济武 Student: LIN Jiwu

大学生健身中心改扩建设计
RECONSTRUCTION AND EXPANSION DESIGN OF COLLEGE STUDENT FITNESS CENTER

傅筱　钟华颖　王铠
FU Xiao, ZHONG Huaying, WANG Kai

课程缘起

大学生健身中心改扩建是南大建筑本科新开设的一个三年级设计课程。在南大过去的设计教学中一直缺少一个专项训练中大跨结构的设计课程，仅是将其融入三年级商业综合体课程设计中。学生既要解决商业综合体的复杂城市关系和复合功能空间，又要研究其中的大跨度设计，其难度难以想象。大学生健身中心改扩建课程设计开设的初衷十分明确，即是将中大跨度空间从商业综合体中分离出来进行分项训练。

选题思考：弱规模，强认知

在中大跨度建筑设计训练方面，国内许多建筑院校做出了值得学习的探索，他们多数是以中大型体育馆、演艺中心或者文化中心为授课载体，以结构选型为基础，结合设备训练学生对大型建筑场馆的综合设计能力。有的院校将结构设计独立成8周专项训练，然后再结合进8周的综合场馆设计。这些训练通常放置在四年级或者毕业设计中，不少院校设计的周期长达16周，让学生得到较为充分的训练。对于南大"2+2+2"的建筑教育模式而言，是不适宜进行长周期训练的。在短短的8周教学时间内，我们应该让学生建立何种认知和培养何种能力是关键。教学团队意识到高校课程设计并非实际工程训练，与漫长的职业生涯相比，学生的认知能力比实际操作能力更为重要，设计的规模和技术复杂并不一定能提升他们对设计本质的理解。因此，结合南大自身的教学特征，教学团队在选题上做了如下思考：

将结构训练的重点放在结构、形态与空间的关联性上。在以前的课程设计中学生较为熟悉的是框架、砖混结构，虽然每一次课程设计均与结构不可分割，但是结构被外围保护材料包裹，在建筑表达上始终处于被动"配合"的状态。对于有一定跨度的空间，容易让学生建立一种"主动"运用结构的意识。为了达到这个训练目标，教学团队认为跨度不宜太大，以30m左右为宜，结构选型余地较大，结构与空间的配合受到技术条件的制约相对较小，以利于学生充分理解空间语言与结构的关联。

强调建筑设计训练的综合性，结构只是一个重要的要素，即不过分放大结构的作用，在将结构作为空间生成的一种推动力的同时要求学生综合考虑场地、使用、空间感受、采光和通风等基本要素。实际上通过短短8周的教学就希望学生能够达到实际操作层面的综合性是不现实的。教学的目标是让学生建立起综合性的认知，所以题目设置宜简化使用功能，明确物理性能要求，给定设备和辅助空间要求，将琐碎的知识点化为知识模块给定学生，让学生更多地学会模块间的组织，而不需要确切了解其中的每一个技术细节。技术细节将随着学生今后的职业生涯而不断增长。技术细节在今天是日新月异的，在高校教授的过多的技术细节或者教师的所谓工程经验，也许在学生毕业时就已经面临淘汰。所以在高校应教授"认知"，进而教授由正确认知带来的技术方法，最终升华为学生的设计哲学。因此，在具体的教学中，结构选型是指定的，但鼓励学生结合性能需求进行合理改变；采光、通风要求是明确的，但鼓励学生结合人的需求进行设计；设备和辅助空间是给定的，但要求学生学会在空间上进行合理布置，理解只有设备和辅助空间布置妥当才能创造使用空间的价值。

教学方法：从"一对一"到"一对多"

本科建筑设计教学方法通常采用教师一对一改图模式，这种模式的优点是教师易于将设计建议传递给学生，当学生不理解时，还可以方便地用草图演示给学生，这适合于手绘时代的交流，甚至教师帅气的草图也是激励学生进步的一种手段。其缺点是学生的问题和教师的建议均局限在一对一的情景中，不能让更多的学生受益，如果教师发现共性问题需临时召集学生再行讲解，其时效性、生动性都大打折扣。鉴于此，我们采用了"一对多"的改图模式，教师和学生全部围坐在投影仪前面，由每个学生讲解自己的设计，教师做点评，并同时鼓励其他学生发表点评意见，当教师认为需要辅助草图说明问题时，教师采用Pad绘制草图，并让每个学生都能看见。

从教学效果看，一对多的改图模式的优点是明显的。首先这样的改图方式具有较好的课堂氛围，学生面对大屏幕讲解自己的设计增强了课堂的仪式感，仪式感的背后是对学习的敬畏之心。同时因为教师不是站在高高的讲台上，又具有平等讨论的轻松气氛。其次，学生既体验到当众讲解的压力，也体验到因出色的设计呈现带来的喜悦，因而无形中激发了他们用功的积极性，而且有利于教师发现学生的共性问题，避免同一问题反复讲却反复犯的教学通病，大大提高了课堂效率。

Course origin

The reconstruction and expansion of college student fitness center is a new third grade design course for architectural undergraduates of Nanjing University. In the past design teaching of NJU, there was a lack of a design course of long-span structure in special training, and it was integrated into the curriculum design of commercial complex in grade 3. Students should not only solve the complex urban relations and composite functional space of commercial complex, but also study the long-span design, which is difficult to imagine. The original intention of the course design for the reconstruction and expansion of college student fitness center is very clear, that is, to separate the medium and large-span space from the commercial complex for itemized training.

Topic selection: weak scale, strong cognition

In terms of medium and large-span architectural design training, many domestic architectural colleges and universities have made explorations worthy of learning. Most of them take medium and large-scale gymnasiums, performing arts centers or cultural centers as the teaching carrier, based on structural selection and combined with equipment to train students' comprehensive design ability of large-scale architectural venues. Some institutions separate the structure into 8-week special training, and then integrate it into the design of 8-week comprehensive venues. These trainings are usually placed in the

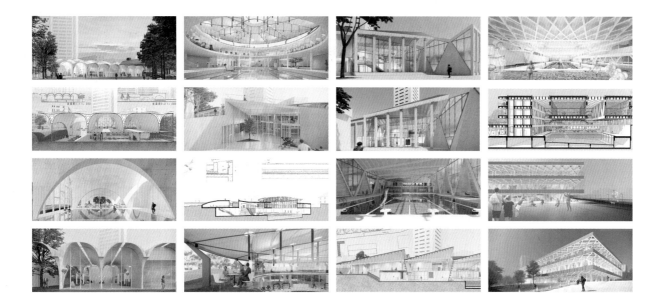

fourth grade or graduation project. The design cycle of many colleges and universities is as long as 16 weeks, so that students can get more adequate training. For the "2 + 2 + 2" architectural education model of Nanjing University, it is not suitable for long-term training. In the short 8-week teaching time, the key is what kind of cognition and ability we should let students establish. The teaching team realizes that college curriculum design is not practical engineering training. Compared with a long career, students' cognitive ability is more important than practical operation ability. The large scale of design and the complexity of technology do not necessarily improve their understanding of the essence of design. Therefore, combined with the teaching characteristics of NJU, the teaching team has made the following reflections on the topic selection:

The focus of structural training is on the relationship between structure, form and space. In the previous curriculum design, students were familiar with the frame and brick concrete structure. Although each curriculum design was inseparable from the structure, the structure was wrapped by external enclosure materials and was always in a passive "cooperation" state in architectural expression. For a space with a certain span, it is easy for students to establish a sense of "active" use of the structure. In order to achieve this training goal, the teaching team believes that the span should not be too large. It is appropriate to be about 30 m. There is a large room for structural selection, and the coordination between structure and space is relatively less restricted by technical conditions, so as to help students fully understand the relationship between spatial language and the structure.

It emphasizes the comprehensiveness of architectural design training, and the structure is only an important element. That is to say, the role of structure is not enlarged. While taking the structure as a driving force for space generation, students are required to comprehensively consider the basic elements such as the site, use, space feeling, daylighting and ventilation. In fact, it is unrealistic to hope that students can achieve the comprehensiveness of practical operation through just 8 weeks of teaching. The goal of teaching is to enable students to establish a comprehensive cognition. Therefore, the topic setting should simplify the use function, clarify the requirements of physical performance, offer equipment and auxiliary space requirements, and turn trivial knowledge points into knowledge modules. Students learn more about the organization between modules without knowing every technical detail. Technical details will continue to grow with students' future career. Today, technical details are changing with each passing day. Too many technical details or teachers' so-called engineering experience taught in colleges and universities may face elimination when students graduate. Therefore, we should teach "cognition" in colleges and universities, then teach the technical methods brought by correct cognition, and finally promote students' design philosophy. Therefore, in specific teaching, structure selection is specified, but students are encouraged to make reasonable changes in combination with performance requirements; the requirements for daylighting and ventilation are clear, but students are encouraged to design according to people's needs; the equipment and auxiliary space are given, but students are required to learn to arrange the space reasonably and understand that only the equipment and auxiliary space are arranged properly can create the value of using the space.

Teaching method: from one-to-one to one-to-many

The undergraduate architectural design teaching method usually adopts the one-to-one mode of modifying pictures by teachers. The advantage of this mode is that teachers can easily pass design suggestions to students. When students do not understand, they can also easily demonstrate to students with sketches, which is suitable for communication in the hand-drawn era, and even the teacher's handsome sketches are a means of motivating students to progress. The disadvantage is that students' problems and teachers' suggestions are limited to one-to-one situations, which cannot benefit more students. If teachers find common problems and need to call students to explain them again, the timeliness and vividness are greatly reduced. In view of this, we adopt the one-to-many mode of modifying pictures. Teachers and students all sit in front of the projector, each student explaines their own design, the teacher makes comments, and at the same time encourages other students to make comments. When the teacher thinks that it is necessary to assist the sketch to explain the problem, the teacher uses the Pad to draw the sketch and let every student see it. From the perspective of teaching effect, the advantages of the one-to-many modifying pictures mode are obvious. First of all, this way of modifying pictures has a good classroom atmosphere. Students explain their design to the big screen, which enhances the sense of ceremony in the classroom. Behind it is the reverence for learning, and at the same time, because teachers are not standing on a high platform, they also have a relaxed atmosphere of equal discussion. Secondly, students not only experience the pressure of public explanation, but also experience the joy brought by excellent design presentation, which invisibly stimulates their enthusiasm for working hard; in addition, it is helpful for teachers to discover common problems of students and avoid the common teaching problem of repeating the same problem but repeating it, greatly improving the efficiency of the classroom.

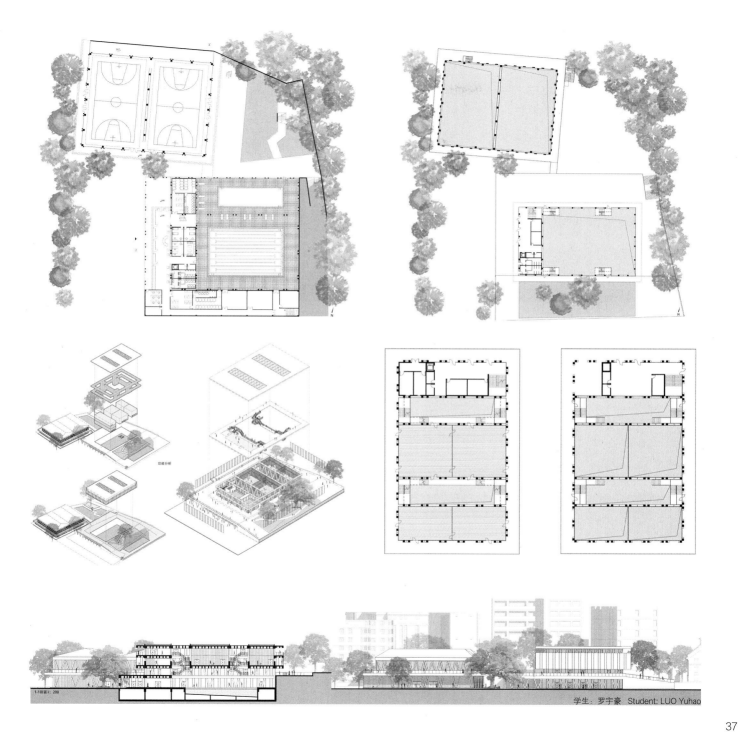

学生：罗宇豪　Student: LUO Yuhao

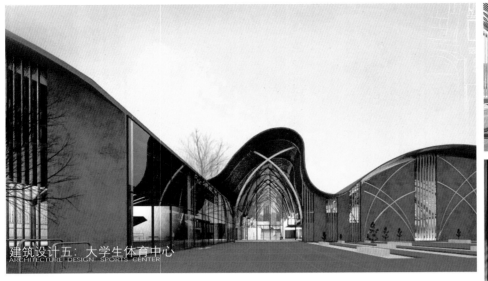

建筑设计五：大学生体育中心
ARCHITECTURE DESIGN: SPORTS CENTER

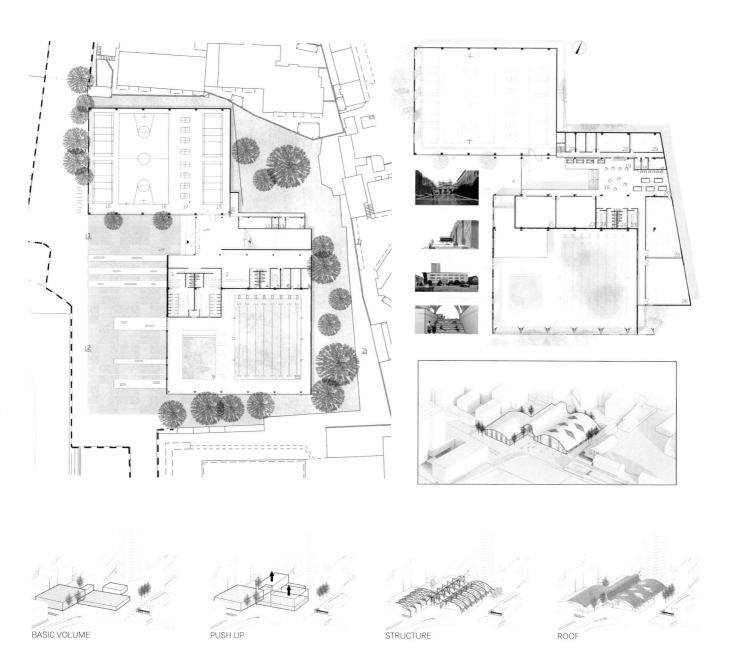

BASIC VOLUME   PUSH UP   STRUCTURE   ROOF

学生：张新雨   Student: ZHANG Xinyu

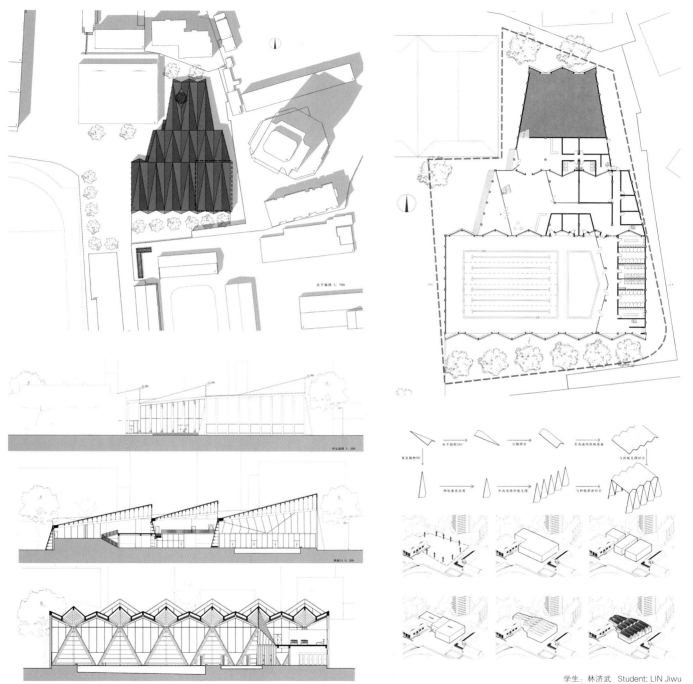

学生：林济武　Student: LIN Jiwu

社区文化艺术中心设计
DESIGN OF COMMUNITY CULTURE AND ART CENTER

张雷 钟华颖 王铠
ZHANG Lei, ZHONG Huaying, WANG Kai

概况

本项目拟在百子亭历史风貌区基地处新建社区文化中心，项目不仅为周边居民提供文化基础设施，同时也期望成为复兴老城的街区活力的文化地标。根据基地条件、功能使用进行建筑和场地设计，总建筑面积约为8 000 m²。总用地详见附图，基地用地面积为4 600 m²。民国时期，百子亭一带属于高级住宅区，在位置上紧临作为文教区的鼓楼以及作为市级行政区的傅厚岗地区。凭借区域上的优势与政府的扶持，百子亭一带自1930年代开始，逐渐成为当时文化精英、社会名流与政府要员的聚居之地。众多受邀来南京创建其事业的学者、文人都在此购买土地，并建造出了一幢幢"和而不同"的新式住宅。这些建筑既是近代南京城市肌理中的现代图景，也是当时中国有为之士们实践其梦想的舞台，还是中国近现代建筑史中不可忽视的华美段落。

基地条件：现状

根据《南京历史文化名城保护规划（2010—2020）》，百子亭历史风貌区被列入"保护名录"，历史风貌区内现有市级文物保护单位3处，为桂永清公馆旧址、徐悲鸿故居和傅抱石故居，不可移动文物8处，历史建筑1处。

教学内容

1）演艺中心

包含400座小剧场，乙级。台口尺寸12 m×7 m。根据设计的等级确定前厅、休息厅、观众厅、舞台等面积。观众厅主要为小型话剧及戏剧表演而设置。按60—80人化妆布置化妆室及服装道具室，并设2—4间小化妆室。要求有合理的舞台及后台布置，应设有排练厅、休息室、候场区以及道具存放间等设施，其余根据需要自定。

2）文化中心

定位为区级综合性文化站，包括公共图书阅览室、电子阅览室、多功能厅、排练厅以及辅导培训、书画创作等功能室（不少于8个且每个功能室面积应不低于30 m²）。

3）配套商业

包含社区商业以及小型文创主题商业单元。其中社区商业为不小于200 m²超市一处，文创主题商业单元面积为60—200 m²。

4）其他

变电间、配电间、空调机房、售票、办公、厕所等服务设施根据相关设计规范确定，各个功能区可单独设置，也可统一考虑。地上不考虑机动车停车配建，街区地下统一解决，但需要根据建筑功能面积计算数量。

教学成果

每人不少于4张A1图纸，图纸内容包括：

1）城市与环境：总平面图1∶500，总体鸟瞰图、轴测图。

2）空间基本表达：平、立、剖面图1∶200—1∶400。

3）空间解析与表现：概念分析图、空间构成分析图、轴测分析图、剖透视图（不少于2张，必须包含大空间、公共空间的剖透视图）、室内外人眼透视图若干。

4）手工模型：每个指导教师组做一个1∶500总平面图体量模型，每位学生做一个1∶500的概念体块模型。

教学进度

本次设计课程共8周。

第一周：授课（1学时）、调研场地及案例、制作场地模型（SU模型＋实体模型1∶500）相应的案例资料收集。

第二周：学生收集案例汇报、初步概念方案讨论（包含体块与场地关系布局、内部空间基本布局）。

第三周：概念深化，完善初步建筑功能布局和空间形态方案（包括基本空间单元及其组合），制作空间形态模型。

第四周：方案定稿，明确空间表皮、平面功能、街区环境模式。

第五周：方案深化Ⅰ，空间表皮、平面功能、街区环境深化。

第六周：方案深化Ⅱ，细化表皮处理、剧场空间及其他重要公共空间的设计。

第七周：方案表达，完成平、立、剖面图绘制，完善SU设计模型。

第八周：制图、排版。

Overview

The project plans to build a new community culture and art center at the base of Baiziting historical area, with a total construction area of about 8 000 m². The project not only serves the cultural infrastructure of surrounding residents, but also hopes to become a cultural landmark to revive the vitality of the old city. Design the building and site according to the base conditions and functional use. The total land is shown in the attached figure, and the land area of the base is 4 600 m². During the period of the Republic of China, Baiziting was a high-grade residential area, close to the Drum Tower as a cultural and educational area and Fuhougang area as a municipal administrative area. With regional advantages and government support, Baiziting area gradually became a settlements for cultural elites, celebrities and government officials since the 1930s. Many scholars invited to Nanjing to establish their careers bought land here and built new houses of "harmony but difference". These buildings

are not only the modern picture in the modern urban texture of Nanjing, but also the stage for Chinese promising people to practice their dreams at that time, and the gorgeous paragraphs that can not be ignored in the history of Chinese modern architecture.

Base conditions: current situation
According to the *Plan for the Protection of Famous Historical and Cultural Cities in Nanjing* (2010–2020), Baizitng historical area is included in the "protection list". There are 3 municipal cultural relics protection units in the historical area, including the former site of GUI Yongqing residence, Xu Beihong's former residence and Fu Baoshi's former residence, 8 immovable cultural relics and 1 historical building.

Teaching content
1) Performing arts center
It contains 400 small theatres, class B. The size of the proscenium is 12 m × 7 m. Determine the area of front hall, lounge, auditorium and stage according to the design level. The auditorium is mainly set up for small-scale drama and drama performances. The dressing room and clothing props room shall be arranged for 60–80 people, and 2–4 small dressing rooms shall be set. Reasonable stage and backstage layout is required. Rehearsal hall, lounge, waiting area, props storage room and other facilities shall be set, and the rest shall be determined according to needs.

2) Cultural center
Located at the district level comprehensive cultural station, it includes public book reading room, electronic reading room, multi-functional hall, rehearsal hall, counseling and training, calligraphy and painting creation and other functional rooms (no less than 8, and the area of each functional room shall not be less than 30 m^2).

3) Supporting business
It includes community business and small cultural and creative theme business units. Among them, the community business is a supermarket with an area of no less than 200 m^2, and the area of cultural and creative theme business unit is 60–200 m^2.

4) Others
Service facilities such as substation room, power distribution room, air conditioning room, ticketing, office and toilet are determined according to relevant design specifications. Each functional area can be set separately or considered uniformly. The parking allocation of motor vehicles is not considered on the ground, and the underground of the block is solved uniformly, but the quantity needs to be calculated according to the building functional area.

Teaching achievements
At least 4 A1 drawings per person, including:
1) City and environment: General layout 1:500, overall aerial view, axonometric drawing.
2) Basic spatial expression: Plan, elevation and section 1:200–1:400.
3) Spatial analysis and expression: Conceptual analysis diagram, spatial composition analysis diagram, axonometric analysis diagram, sectional perspective view (no less than 2, which must include the sectional perspective view of large space and public space), indoor and outdoor human eye perspective view.
4) Manual model: Each tutor group makes a 1:500 general layout volume model, and each student makes a 1:500 concept block model.

Teaching progress
This design course is 8 weeks in total.
Week 1: Teaching (1 h), researching sites and cases, making site models (SU model + physical model 1:500) and collecting relevant case data.
Week 2: Students collect case reports and discuss preliminary conceptual plans (including the layout of the relationship between the block and the site, and the basic layout of the interior space).
Week 3: Concept deepening, perfecting the preliminary architectural functional layout and spatial form scheme (including basic spatial units and their combinations), and making a spatial form model.
Week 4: The plan is finalized, and the space skin, plane function, and block environment pattern are defined.
Week 5: Plan deepening I, the space skin, plane function, and block environment are deepened.
Week 6: Plan deepening II, refining of skin treatment, design of theater space and other important public spaces.
Week 7: Schematic expression, completing the plan, elevation and section, and improving SU design model.
Week 8: Drawing, typesetting.

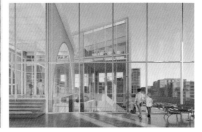

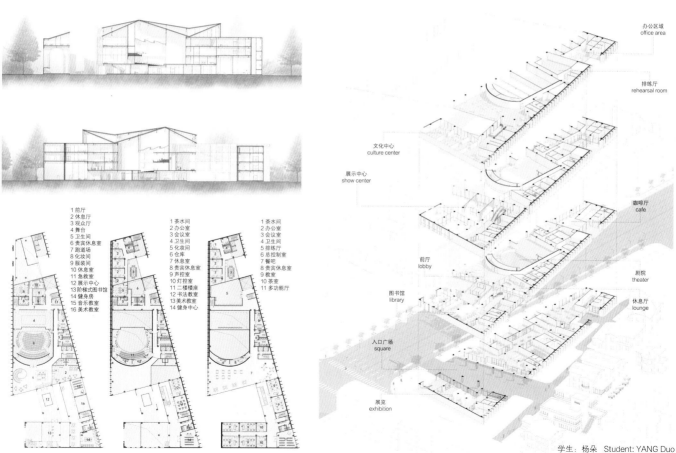

学生：杨朵　Student: YANG Duo

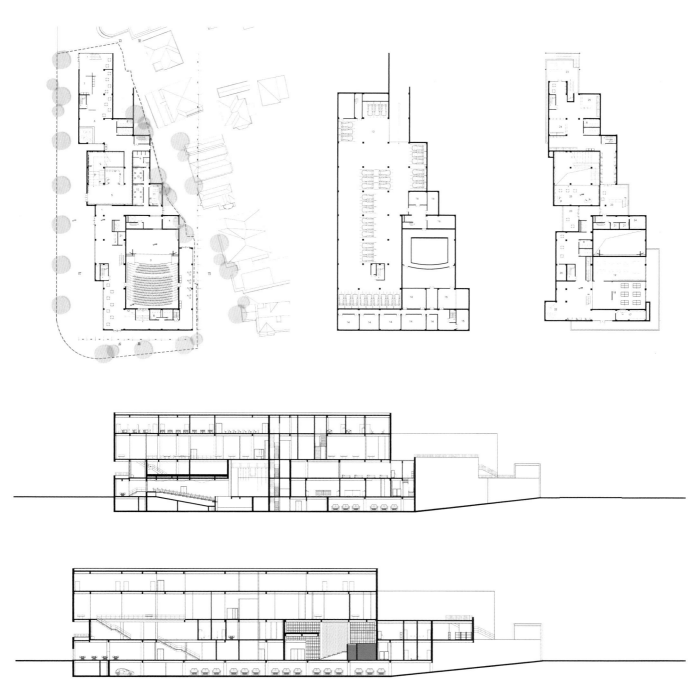

学生：田舒琳　Student: TIAN Shulin

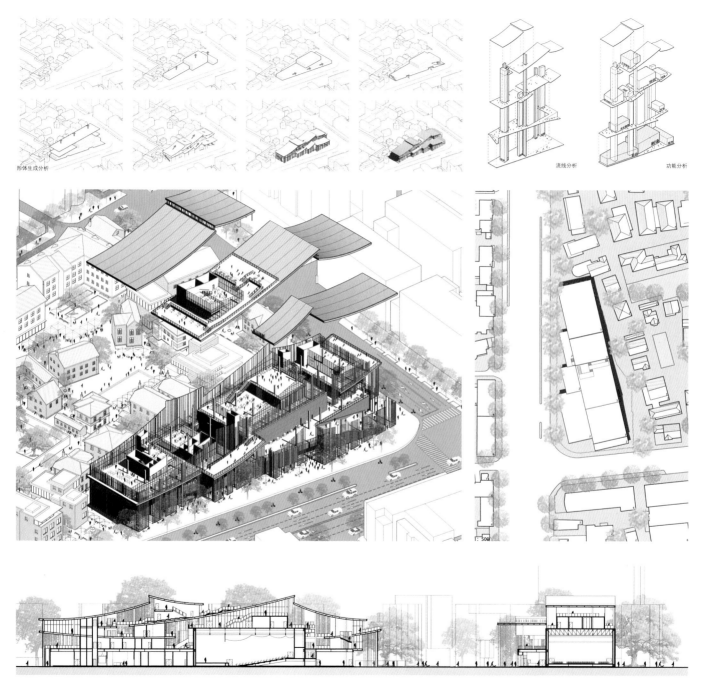

形体生成分析　　　　　流线分析　　　功能分析

学生：罗宇豪　Student: LUO Yuhao

高层办公楼设计
DESIGN OF HIGH-RISE OFFICE BUILDINGS

吉国华　胡友培　尹航
JI Guohua, HU Youpei, YIN Hang

教学目标

　　生态性能驱动的办公建筑设计涉及城市、空间、形体、环境、能耗、结构、设备、材料、消防等方面内容，是一项较复杂与综合的任务。有效的空间组织、适应性形体、交互性表皮以及性能化结构设计等策略，对建筑室内外环境的生态性能起着决定性的作用。本课题教学重点和目标是帮助学生理解、消化以上知识，提高综合运用并创造性解决问题的技能，学习并运用生态性能模拟分析软件，以生态性能驱动建筑设计。

设计条件与要求

　　1）经济技术指标与场地

　　用地面积为 4 520 m^2，地上总建筑面积 ≥ 35 000 m^2，建筑限高 ≤ 100 m。

　　2）功能要求

　　办公：设计应兼顾各种办公空间形式。

　　会议：须设置 400 人报告厅 1 个、200 人报告厅 2 个、100 人报告厅 4 个，其他各种会议形式的中小型会议室若干，以及咖啡/茶室、休息厅、服务用房等。

　　机动车交通：机动车交通独立设置，人车分离。场地交通流线，须结合现状周边情况，统一考虑。地下部分为车库和设备用房。

　　3）相关规范

　　《民用建筑设计通则》（GB 50352—2005）；

　　《城市道路和建筑物无障碍设计规范》（JGJ 50—2001）；

　　《办公建筑设计规范》（JGJ 67—2006）；

　　《车库建筑设计规范》（JGJ 100—2015）；

　　《建筑设计防火规范》（GB 50016—2014）；

　　《汽车库、修车库、停车场设计防火规范》（GB 50067—2014）。

　　4）其他

　　用地红线及建筑退让线详见总平面图。汽车库和自行车库的配置应满足《南京市建筑物配建停车设施设置标准与准则》的要求。

　　底层的架空层面积不计入建筑总面积。

教学进度（8 周）

　　第一周：场地调研与分析、案例研究、初步概念。

　　第二周：概念深化。

　　第三、四周：总平面设计（草图、1∶500 草模）。

　　第五、六、七周：平面、立面与细部深化设计。

　　第八周：成果表达与制作。

成果要求

　　建筑总平面图（1∶500）。

　　建筑平、立、剖面图（1∶200）。

　　建筑大样图（≥ 1∶20）。

　　建筑表现图若干。

　　建筑形体研究模型（1∶500）。

　　建筑模型（1∶200/1∶300）。

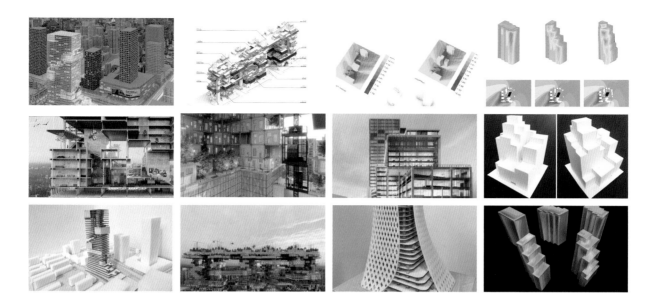

Teaching objectives

The design of office buildings driven by eco-performance involves the aspects of the city, space, form, environment, energy consumption, structure, equipment, materials, and fire protection. It is a complex and comprehensive task. The strategies such as effective spatial organization, adaptive shapes, interactive surfaces, and performance-based structural design play a decisive role in ecological performance of indoor and outdoor environment. This course intends to help the students to understand and digest the knowledge of various aspects, improve comprehensive application and creative problem solving skills, learn and use the ecological performance simulation analysis software, and drive architectural design with ecological performance.

Design conditions and requirements

1) Economic and technical indicators and site

Land area: 4 520 m², the total building area above ground 35 000 m², height limit ≤ 100 m.

2) Functional requirements

Office: The design should take into account various forms of office space.

Meeting: There should be a 400-person conference hall, two 200-person conference halls, and four 100-person conference halls. There should also be several small and medium-sized conference rooms in various forms, and the cafe/tea bar, lounge, and service room.

Vehicle traffic: The vehicle traffic should be set independently, with separation between people and vehicles. Site traffic flow should be considered in combination with the surrounding situations. The underground part consists of the garage and equipment room.

3) The relevant specifications

Code for Design of Civil Buildings (GB 50352—2005);
Code for Design on Accessibility of Urban Road and Buildings (JGJ 50—2001);
Design Code for Office Building (JGJ 67—2006);
Code for Design Garage Buildings (JGJ 100—2015);
Code for Fire Protection Design of Buildings (GB 50016—2014);
Code for Fire Protection Design of Garage, Motor-Repair-Shop and Parking-Area (GB 50067—2014).

4) Others

See the red line of land use and building yield line in the general layout. The configuration of the garage and bicycle garage should meet the requirements of the *Standards and Guidelines for Setting Parking Facilities for Buildings in Nanjing*. The area of the raised floor is not included in the total building area.

Teaching progress (8 weeks)

Week 1: Site investigation and analysis, case research, initial concept.
Week 2: Concept deepening.
Weeks 3 & 4: General layout design (sketch, draft model – 1:500).
Weeks 5, 6 & 7: Plan, elevation and detail deepening design.
Week 8: Achievement expression and preparation.

Achievement requirements

Architectural general layout (1:500).
Architectural plan, elevation, and section (1:200).
Detail drawing (≥ 1:20).
Several architectural performance drawings.
Building form research model (1:500).
Building model (1:200/1:300).

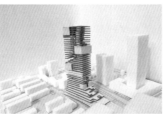

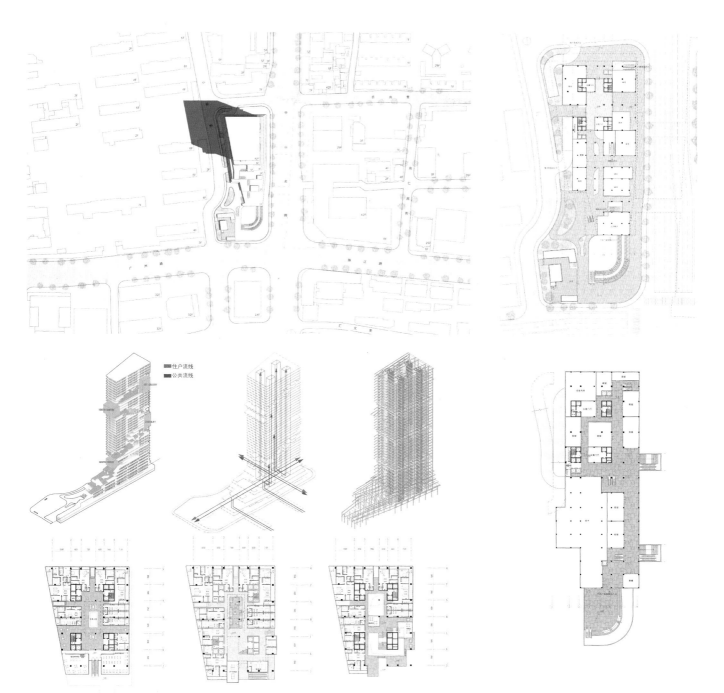

学生：陈露茜 顾天奕　Student: CHEN Luqian，GU Tianyi

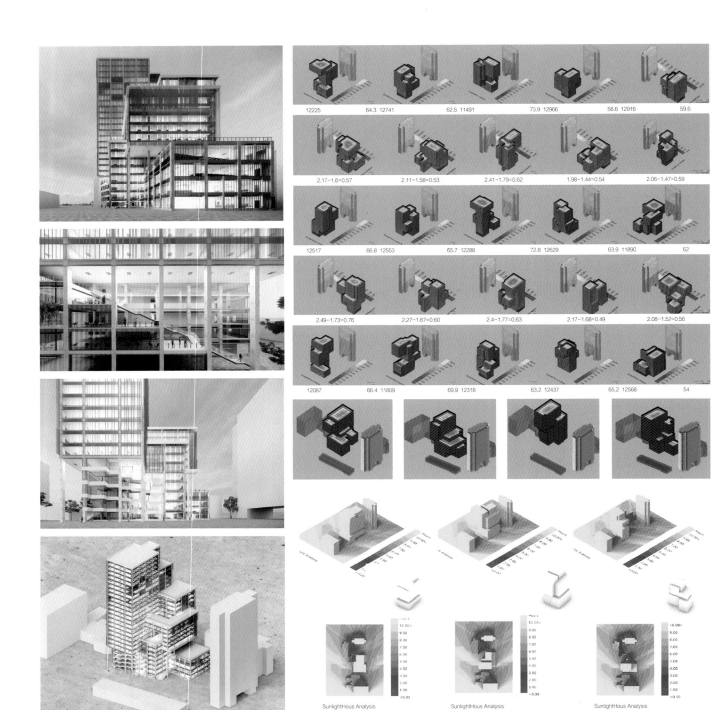

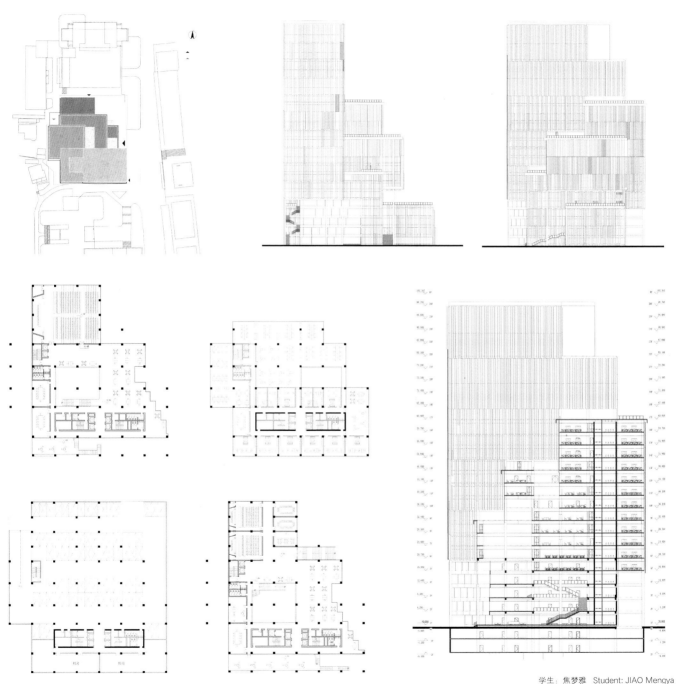

学生：焦梦雅 Student: JIAO Mengya

城市设计
URBAN DESIGN
童滋雨　尹航　尤伟
TONG Ziyu, YIN Hang, YOU Wei

课程内容
　　计算化城市设计

教学目标
　　中国的城市发展已经逐渐从增量扩张转向存量更新。通过对城市建成环境的更新改造而提升环境性能和质量，将成为城市建设的新热点和新常态。与此同时，5G、物联网、无人驾驶等技术的发展又给城市环境的使用方式带来了新的变化。如何在城市更新设计中拓展建筑设计的边界也就成为新的挑战。
　　城市更新不但需要对建成环境本身有更充分的认知，也要对其中的人流、车流乃至水流、气流等各种动态的活动有正确的认知。从设计上来说，这也大大提高了设计者所面临的问题的复杂性，仅靠个人的直观感受和形式操作难以保证设计的合理性。而借助空间分析、数据统计、算法设计等数字技术，我们可以更好地认知城市形态的特征，理解城市运行的规则，并预测城市未来的发展。通过规则和算法来计算生成城市也是对城市设计思维范式的重要突破。
　　因此，本次设计将针对这些发展趋势，以城市街巷空间为研究对象，通过思考和推演探索其更新改造的可能性。通过本次设计，学生们可以理解城市设计的相关理论和方法，掌握分析城市形态和创造更好城市环境质量的方法。

设计场地
　　课程设计范围位于南京市鼓楼区广州路两侧，东至中央路，西至上海路，总长度约750 m，南北边界可根据调研自己确定。

成果要求
　　本次设计以小组为单位，每小组2人。每组成果包括8张A1图纸和1份A4成果文本，具体内容可包括但不限于以下部分。
　　1）设计表达：平面图、立面图、轴测图、透视图等；
　　2）设计推演：设计形成过程的分析图；
　　3）设计评估：对设计成果的各种评估分析。

教学进度
　　阶段一：　场地调研与案例分析（2周）。
　　阶段二：　设计目标确定与总体布局方案（2周）。
　　阶段三：　方案完善与局部深化（3周）。
　　阶段四：　制图与排版（1周）。

Course content
Computerized urban design

Teaching objectives
China's urban development has gradually shifted from incremental expansion to stock renewal. Improving environmental performance and quality through the renewal and transformation of urban built environment will become a new hot spot and new normal of urban construction. At the same time, the development of 5G, Internet of things, unmanned driving and other technologies has brought new changes to the use of urban environment. How to expand the boundary of architectural design in urban renewal design has become a new challenge.

Urban renewal not only needs to have a better understanding of the built environment itself, but also has a correct understanding of the people stream, vehicles stream, water stream, air stream and other dynamic activities. In terms of design, it also greatly improves the complexity of the problems faced by designers. It is difficult to ensure the rationality of design only by personal intuitive feeling and formal operations. With the help of various digital technologies such as spatial analysis, data statistics and algorithm design, we can better understand the characteristics of the urban form, understand the rules of the urban operation, and predict the future development of the city. Calculating and generating cities through rules and algorithms is also an important breakthrough in the thinking paradigm of urban design.

Therefore, this design will aim at these development trends, take the urban street space as the research object, and explore the possibility of its renewal and transformation through thinking and deduction. Through this design, students can understand the relevant theories and methods of urban design, and master the methods of analyzing the urban form and creating better urban environmental quality.

Design site
The course design scope is located on both sides of Guangzhou Road, Gulou District, Nanjing, east to Zhongyang Road and west to Shanghai Road, with a total length of about 750 m. The north-south boundary can be determined according to the investigation.

Achievement requirements
This design takes the group as the unit, with 2 people in each group. Each group of achievements includes 8 A1 drawings and 1 A4 achievement text. The specific contents can include but are not limited to the following parts:
1) Design expression: Plan, elevation, axonometric drawing, perspective view, etc.
2) Design deduction: Analysis diagram of design formation process.
3) Design evaluation: Various evaluation and analysis of design results.

Teaching progress
Stage 1: Site investigation and case analysis (2 weeks).
Stage 2: Determination of design objectives and overall layout scheme (2 weeks).
Stage 3: Scheme improvement and local deepening (3 weeks).
Stage 4: Drawing and typesetting (1 week).

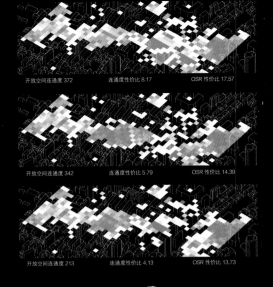

开放空间连通度 372　　连通度性价比 8.17　　OSR 性价比 17.57

开放空间连通度 342　　连通度性价比 5.79　　OSR 性价比 14.39

开放空间连通度 213　　连通度性价比 4.13　　OSR 性价比 13.73

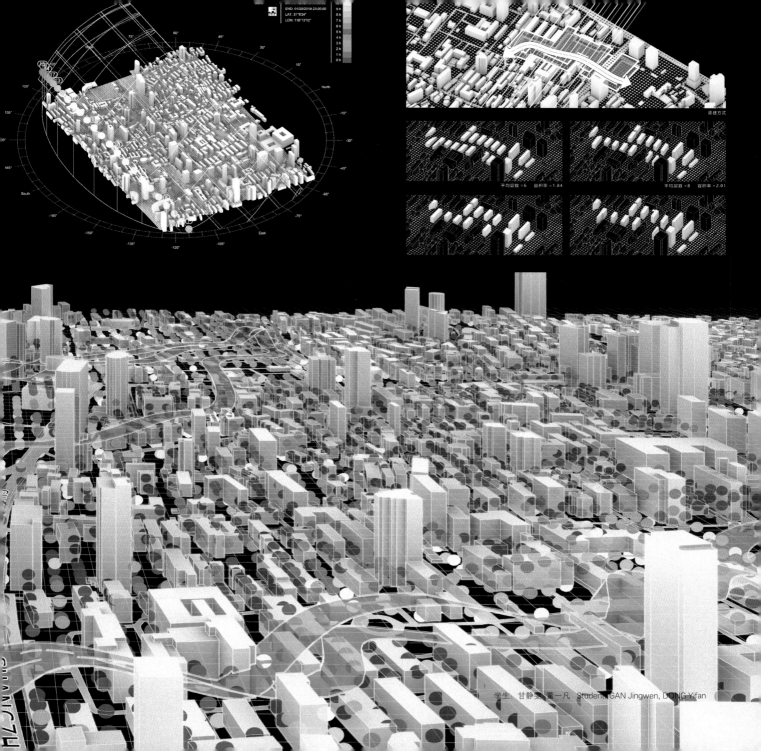

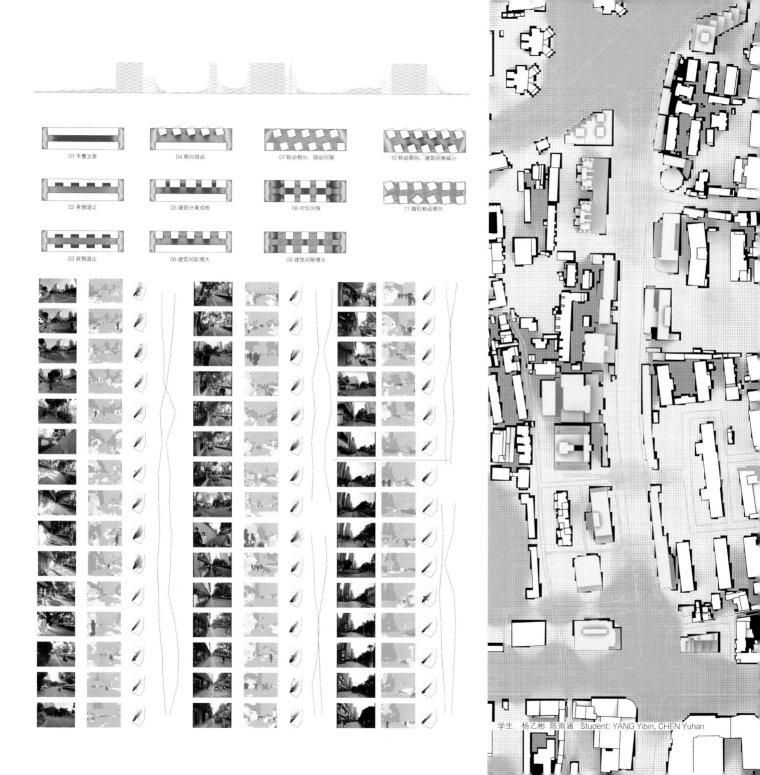

学生：杨乙彬 陈雨涵　Student: YANG Yibin, CHEN Yuhan

基于规则和算法的设计和搭建
DESIGN AND CONSTRUCTION BASED ON RULES AND ALGORITHMS

童滋雨
TONG Ziyu

课程内容
基于规则和算法的设计和搭建

课程介绍
本课题聚焦于设计过程中的规则提取和算法应用,结合参数变量的设置,生成更有趣且合理的设计成果。在此过程中,设计的功能、形态乃至结构都可以是规则提取的目标。而相对于形式本身,我们更关注其潜在规则和算法的科学性。设计包括空间结构体的几何规则生成、参数化生成设计方法和设计算法等。

数字化加工技术的发展更增加了加工和建造的多样性,也为复杂建筑形体的实现提供了可能。目前可以使用的工具包括木工雕刻机、三维打印机和机器人系统。

本课题初步设定是设计一个具有一定跨度和高度的构筑物,该构筑物可通过砌块或编织等方法构建,能容纳3人左右的活动空间。要求空间适宜,结构合理,形体的生成应具有相应的几何规则和算法,并通过数字化加工完成最终的模型搭建。

Course content
Design and construction based on rules and algorithms

Course description
This topic focuses on the rule extraction and algorithm application in the design process, combined with the setting of parameter variables to generate more interesting and reasonable design results. In this process, the function, form and even structure of the design can be the goal of rule extraction. Compared with the form itself, we pay more attention to the scientificity of its potential rules and algorithms. The design includes geometric rule generation, parametric generation design method and design algorithm of the spatial structure.

The development of digital processing technology not only increases the diversity of processing and construction, but also provides the possibility for the realization of complex architectural form. Tools currently available include woodworking engraving machines, three-dimensional printers and robotic systems.

The preliminary setting of this topic is to design a structure with a certain span and height, which can be constructed by means of block or weaving, and can accommodate an activity space for about 3 people. It is required that the space is suitable and the structure is reasonable. The generation of the shape should have corresponding geometric rules and algorithms, and the final model construction should be completed through digital processing.

 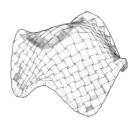 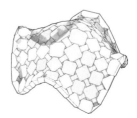 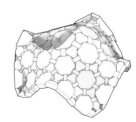

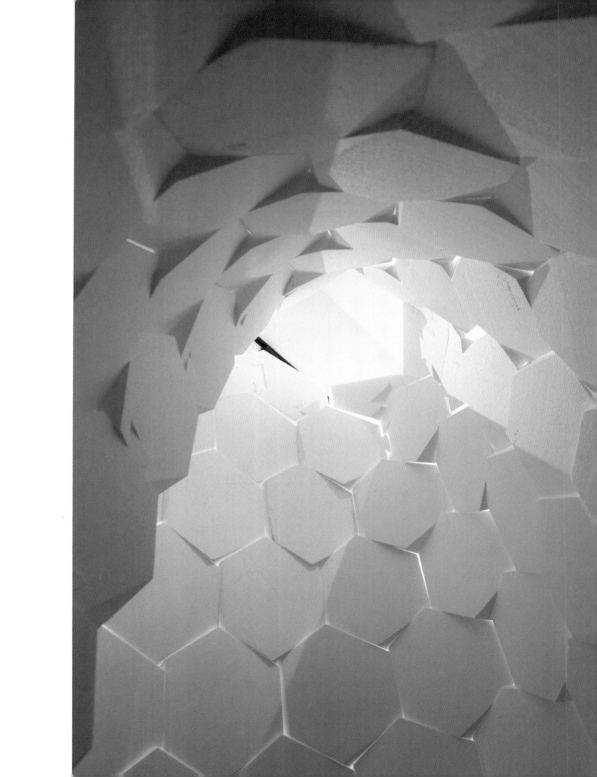

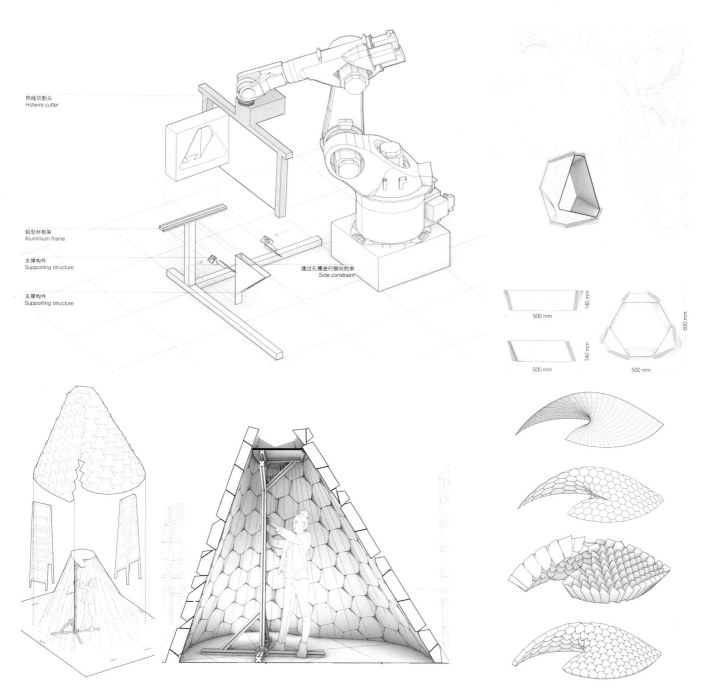

学生：甘静雯 董一凡 陈雨涵　Student: GAN Jingwen, DONG Yifan, CHEN Yuhan

基于力学生形的数字化设计与建造
DIGITAL DESIGN AND CONSTRUCTION ON THE FORM-FROM-FORCE PRINCIPLE
李清朋　吉国华
LI Qingpeng, JI Guohua

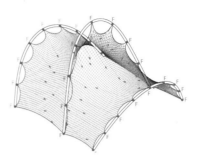

教学目标
本毕业设计基于建筑数字化技术，涵盖案例分析、设计研究以及建造实践三个部分，建立基于力学生形的设计方法，解决数字化设计与实际建造的真实问题，完成从形态设计到数字化建造的全过程。整个课程以结构性能为形态设计的出发点，协同思考形式美学与建造逻辑的关系，培养学生在建筑设计阶段主动考虑结构逻辑的能力，在建筑形式创新和结构逻辑之间寻求统一。

课程介绍
"建筑与结构的关系"是建筑学与建筑设计中最基本、最核心的问题之一。建筑与结构在空间的围合、形体的构筑、形象的塑造等三个方面具有密不可分的关系。从力的感知到受力体系的选择，从结构骨架的支撑到空间形态的实现，从空间形态到建筑作为人类生活空间的容器，基于力学原理的形态设计为建筑空间的设计提供了一个有力的切入点，大大地延伸了建筑与结构协同的操作范围，同时也提供了一个完成设计到建造的起点。

本课题要求学生在学校自选环境中设计一处用地面积 4 m × 4 m、遮盖面积为 10 m^2 左右的建筑空间，以满足师生停留、休憩、交流的功能需求。课题通过实物模型制作不断探索设计问题。用数字化的方法研究和解决问题，最终通过数控加工的方式实现具有真实细节的构筑物。

重点问题
1）基于计算性设计的技术与思维对建筑问题进行解析。
2）结构性能设计的力学原型分析与应用。
3）Grasshopper 程序及编程学习，运用各种程序方法和各类库文件。
4）材料研究，充分挖掘并整理与数控建造相关的各类材料。
5）掌握相关模型制作工具（激光雕刻机、CNC、3D打印机、机械臂等）的基本知识和操作要领。

Taining objectives
Based on building digital technology, this graduation project covers three parts: case analysis, design research and construction practice. It establishes a design method based on the form-from-force principle, solves the real problems between digital design and actual construction, and completes the whole process from form design to digital construction. The whole course takes structural performance as the starting point of form design, cooperatively considers the relationship between formal aesthetics and construction logic, cultivates students' ability to actively consider structural logic in the architectural design stage, and seeks unity between architectural form innovation and structural logic.

Course description
"The relationship between architecture and structure" is one of the most basic and core problems in architecture and architectural design. Architecture and structure have an inseparable relationship in three aspects: the enclosure of space, the construction of form and the shaping of image. From the perception of force to the selection of stress system, from the support of structural skeleton to the realization of spatial form, and from spatial form to architecture as the container of human living space, the form design based on mechanical principles provides a powerful entry point for the design of architectural space and greatly extends the operation range of buildings and structure coordination, It also provides a starting point from design to construction.
The topic requires students to design a building space with a land area of 4 m × 4 m, and covering an area of about 10 m^2 in the school's optional environment, so as to meet the functional needs of teachers and students to stay, rest and communication. The topic continues to explore the design problem through the production of the physical models. The problem is studied and solved by digital methods, and finally the structure with real details is realized by NC machining.

Key issues
1) Analyze architectural problems based on the technology and thinking of computational design.
2) Mechanical prototype analysis and application of structural performance design.
3) Grasshopper program and programming learning, using various program methods and various library files.
4) Material research, fully excavate and sort out various materials related to CNC construction.
5) Master the basic knowledge and operation essentials of related model making tools (laser engraving machine, CNC, 3D printer, robotic arm, etc.).

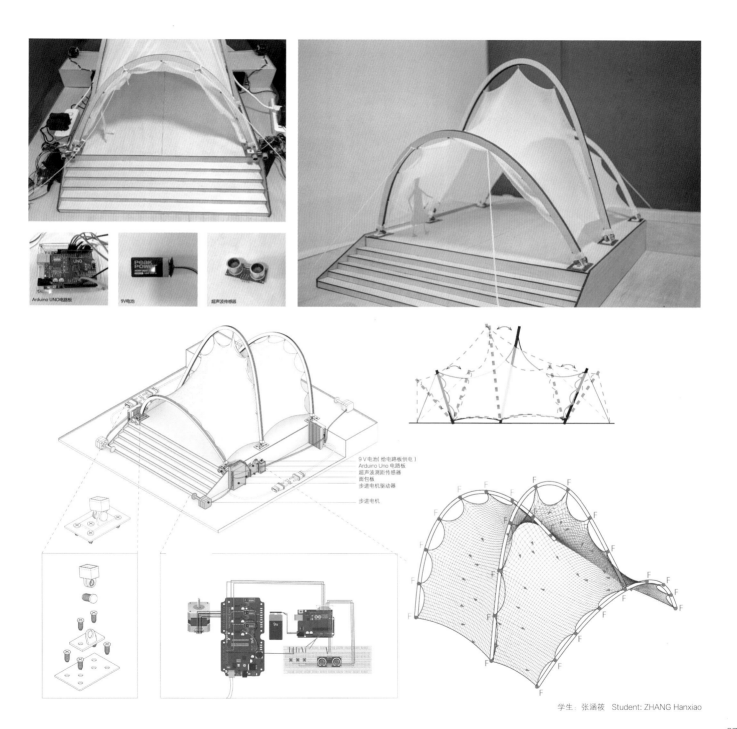

学生：张涵筱　Student: ZHANG Hanxiao

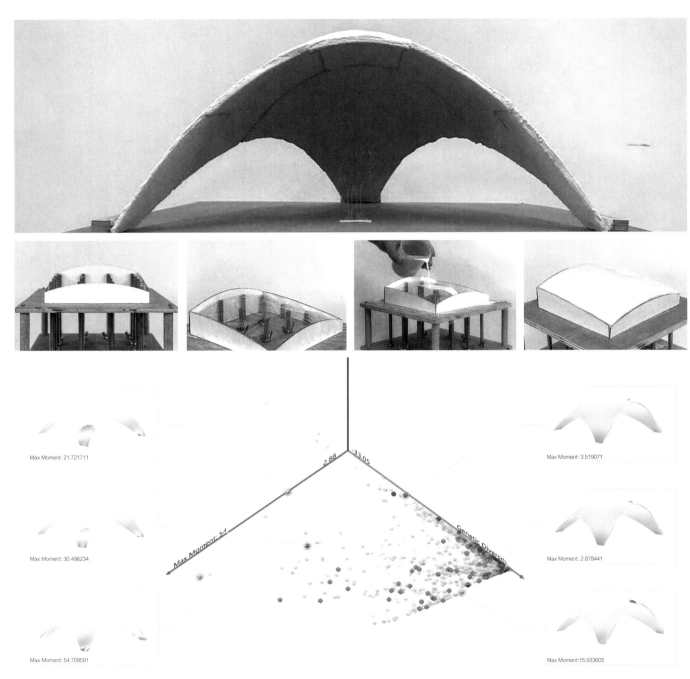

学生：马子昂　Student: MA Zi'ang

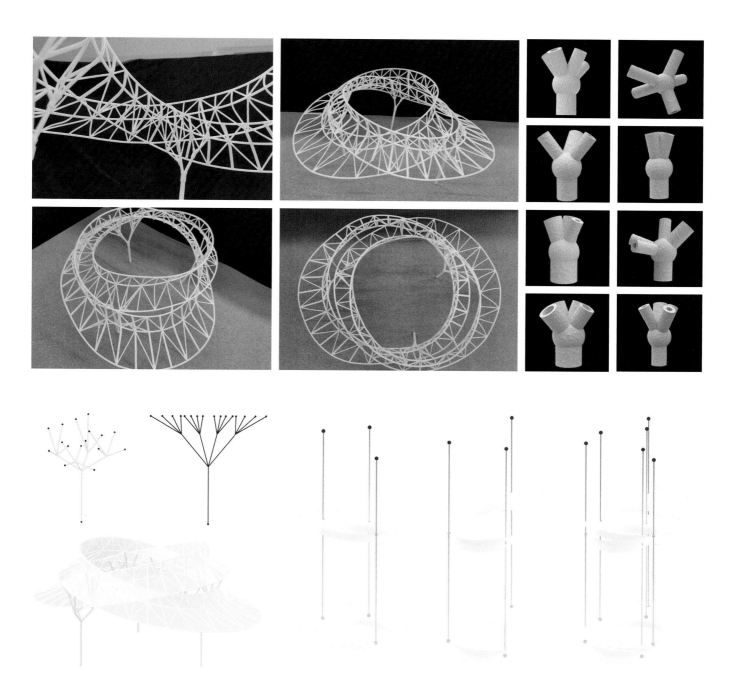

学生：焦梦雅　Student: JIAO Mengya

流动的空间：海上贸易的技术传播与东南亚港口城市的近代塑造
THE SPACE IN MOTION: TRANSMISSION OF TECHNOLOGY ALONG THE MARITIME TRADE ROUTE AND THE MODERN SHAPING OF PORT CITIES IN SOUTHEAST ASIA

黄华青　冉光沛
HUANG Huaqing, RAN Guangpei

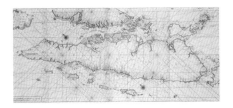

课程介绍

空间的"能动性"是建筑学、人类学等多学科关注的话题，体现于空间形式的自治性和空间—社会的互动建构，进而成为跨越主体与客体、个体与集体、想象与实体研究的桥梁。本研究从列斐伏尔、布迪厄、白馥兰的现代空间理论脉络出发，探讨空间作为一种兼具物质性 / 社会性的"技术"传播媒介，如何作为"能动者"构成并形塑近现代城市聚落的物质和社会景观。

本课题在"一带一路"、聚落文化、近代遗产等视野下，以15—19世纪兴盛的海上贸易沿线港口城市为载体，探讨技术（包括建筑、机械等物质技术及人力、资本、组织等社会技术）的传播与东南亚、南亚近现代港口城市的形成与变迁之间的激烈互动。研究重点包括但不限于巴塔维亚（今印度尼西亚）、马六甲（今马来西亚）、会安（今越南）、马尼拉（今菲律宾）、亚齐（今印度尼西亚）等——这些城市的近现代发展大多受到海上贸易的技术流动以及西方殖民的技术输出的双重驱动，从其城市规划及营建中不难看到来自西方城市理念及贸易需求的影响；而中国作为茶叶、瓷器等生产技术的源头及海上茶路的重要起点，也随着贸易和技术流动而融入这一世界体系的重塑之中，在广州、福州、宁波等沿海港口中可找到诸多线索。

本课题基于史料挖掘和田野调查，从聚落形态和建筑类型出发，结合全球史、社会经济史、社区史等视角，为东南亚近代港口城市的空间形塑寻求以"人"为基准的线索，并探寻中国在这一流动的物质、社会、文化网络中发挥的作用。

教学目标

本课题涉及建筑学、聚落研究、建筑遗产、民族志、移民史、贸易史、技术史等多领域话题，以史料搜集、田野调查、聚落形态及建筑类型研究为手段，成果以毕业论文形式呈现。

Course description

The "agency" of space is a topic concerned by many disciplines such as architecture and anthropology, which is reflected in the autonomy of space form and the interactive construction between space and society, and then becomes a bridge across subject and object, individual and collective, imagination and entity. This study starts from H. Lefebvre, P. Bourdieu and F. Bray's context of modern space theory to discuss how space, as a "technical" media with both material and social characteristics, constitutes and shapes the material and social landscape of modern urban settlements as an "agent".

From the perspective of "one belt, one road", the settlement culture and the modern heritage, also based on the port city that flourished along the 15-19 century, this paper discusses the fierce interaction between the dissemination of technology (including material technology such as construction and machinery, and social technology such as human, capital and organization) and the formation and change of modern port cities in Southeast Asia and South Asia. The research focus includes but is not limited to Batavia (now Indonesia), Malacca (now Malaysia), Hoi'an (now Vietnam), Manila (now Philippine) and Aceh (today's Indonesia), etc.—the modern and contemporary development of these cities is mostly driven by the technology flow of maritime trade and the technology export of western colonies. It is not difficult to see the influence of western urban ideas and trade demand from their urban planning and construction. As the source of tea, porcelain and other production technologies and an important starting point of Maritime Tea Road, China is also driven by trade and technology. Many clues can be found in coastal ports such as Guangzhou, Fuzhou and Ningbo.

Based on historical data mining and field investigation, starting from settlement forms and architectural types, combined with the perspectives of global history, socio-economic history and community history, this topic seeks clues based on "people" for the spatial shaping of modern port cities in Southeast Asia, and explores the role of China in this mobile material, social and cultural network.

Teaching objectives

This subject involves topics in many fields such as architecture, settlement studies, architectural heritage, ethnography, immigration history, trade history, technology history, etc.. It uses historical data collection, fieldwork, settlement form, and architectural type research as means, and the results are presented in the form of a graduation thesis.

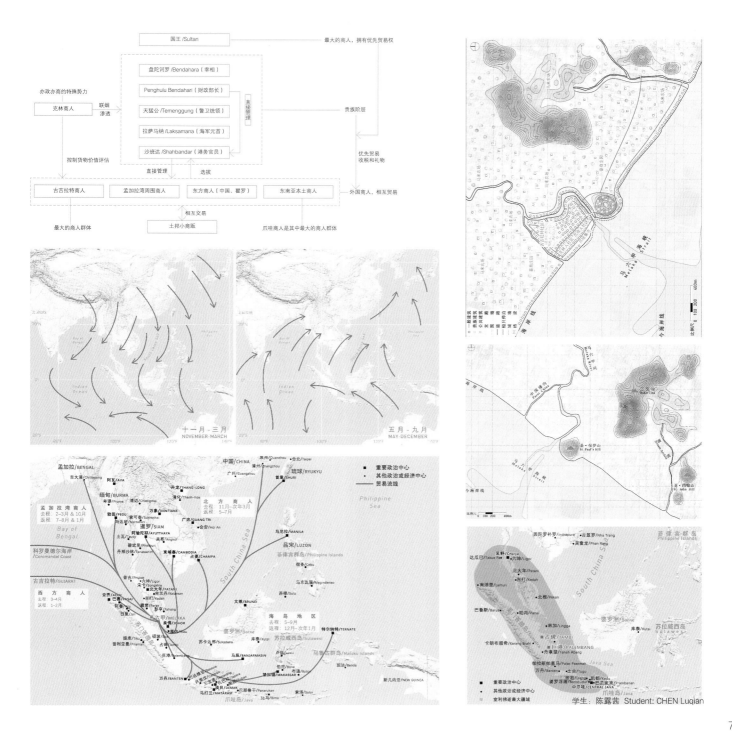

建筑设计研究（一）ARCHITECTURAL DESIGN RESEARCH 1

基本设计
BASIC DESIGN
傅筱
FU Xiao

建筑要求
1）按照南京市宅基地相关规定，建筑基底面积不得超过 130 m²。要求内部布局紧凑经济，使用功能合理，在满足功能需求之下，尽量减少面积以省造价，总建筑面积不得超过 210 m²。
2）可选用结构为：钢筋混凝土框架结构、砖混结构、轻钢龙骨结构体系、木框架结构体系（同一地块的小组不得选用相同的结构体系）。外墙材料选择需与结构体系有一定的关联性，并考虑保温隔热要求。
3）入户空间要求朝南或者朝东。
4）空调形式为分体挂机或柜机，需设计放置位置。
5）明厨明卫。
6）具体房间数量要求：
（1）客厅；（2）餐厅；（3）厨房；（4）客卧 1 间（带卫生间）；（5）主卧 1 间（带卫生间）；（6）次卧 1 间（使用公用卫生间）；（7）画室 1 间（使用面积不小于 30 m²）；（8）公用卫生间，根据需要确定数量；（9）储藏空间。
技术要求：鼓励用 BIM（Revit）、Enscape 设计和表达。
参考书目：《加拿大木框架房屋建筑》（学院资料室）。

教学进度
1）场地分析
时间：第一周
内容：分析场地和相关案例研究，并提交分析报告和 1:100 场地模型（全组一个，用于平时方案讨论，场地范围由学生自行研究确定）。每组分析 2—3 个案例（需包含所选定的结构体系），组与组之间的案例尽量不要重复。
2）了解基本建筑设计原理
时间：第二周
内容：汇报分析报告，并提出初步概念方案，以工作模型或者图纸进行研究，需制作 PPT 汇报。

3）组织空间与行为
时间：第三、四、五、六周
内容：形成解决方案，以工作模型和图纸进行研究，需制作 PPT 汇报。
4）设计研究与表达
时间：第七、八周

成果要求
1）完成平、立、剖面图纸，比例 1:50，平面和剖面要求布置家具、材料填充以及表达人的行为。
2）提交 1:50 场地模型（全组 1 个，用于答辩）和 1:50 单体工作模型（每组 1 个）。
3）表达空间关系的三维剖透视不少于 2 个（不低于 1:50），要求材料填充，必须有家具和人的行为表达。
4）外墙应至少有 2 个详图（不小于 1:20）以表达设计意图。
5）1 张有助于表达形状和空间的透视图，可以是渲染图或模型照片。
6）其他有助于表达设计意图的图纸。

Building requirements
1) In accordance with the relevant regulations of Nanjing City residential base, the building base area shall not exceed 130 m². The internal layout is compact and economical, the usage function is reasonable, and the area is reduced as much as possible to save the construction cost under the satisfaction of the function demand.The total building area shall not exceed 210 m².
2) The optional structures are: reinforced concrete frame structure, brick concrete structure, light steel keel structure system, wood frame structure system (the same plot of the group shall not choose the same structural system). The choice of exterior wall materials should be related to the structural system and should consider the heat insulation requirements.

3) The entrance space should face south or east.
4) The air conditioner is in the form of split hanging machine or cabinet machine, and the placement position should be designed.
5) Bright kitchen and bright bathroom.
6) The number of specific rooms requires that:
(1) Living room; (2) Dining room; (3) Kitchen; (4) 1 guest bedroom (with bathroom); (5) 1 master bedroom (with bathroom); (6) 1 room of secondary bedroom (using common bathroom); (7) 1 drawing room (use area not less than 30 m^2); (8) Common bathroom, the number is determined according to the need; (9) Storage space.
Technical requirements: BIM (Revit), Enscape design and expression are encouraged.
Bibliography: *Canadian Wood Frame House Architecture* (College Reference Room)

Teaching process
1) Site analysis
Time: Week 1
Content: Analyze the site and the related case studies, and submit the analysis report and 1:100 site model (a model for each group for scheme discussion, and the site scope should be determined by the students). Each group should analyze 2–3 cases (including the selected structural system), and the cases between groups should not be repeated.
2) Understand the basic architectural design principle
Time: Week 2
Content: Submit the analysis report, put forward a preliminary conceptual scheme, carry out the research based on the working model or drawing, and prepare a PPT report.
3) Organization space and behaviors
Time: Weeks 3, 4, 5 & 6.
Content: Form a solution, carry out the research based on the working model or drawing, and prepare a PPT report.
4) Design research and expression
Time: Weeks 7 & 8.

Achievment requirements
1) Complete the plan, elevation and section (1 : 50), and the plan and section should involve the layout of furniture, filling of materials and expression of human behaviors.
2) Submit the 1:100 site model (a model for each group for reply) and 1:50 single working model (one for each group).
3) There should be at least 2 three-dimensional sectional perspective views (no less than 1 : 50) that can express the spatial relationship; the filling of materials, layout of furniture, and expression of human behaviors should be involved.
4) There should be at least 2 detail drawings (no less than 1 : 20) of the exterior wall expressing the design intent.
5) There should be a perspective view being conducive to expressing the shape and space, which can be a rendering or model photo.
6) There should be other drawings being conducive to expressing the design intent.

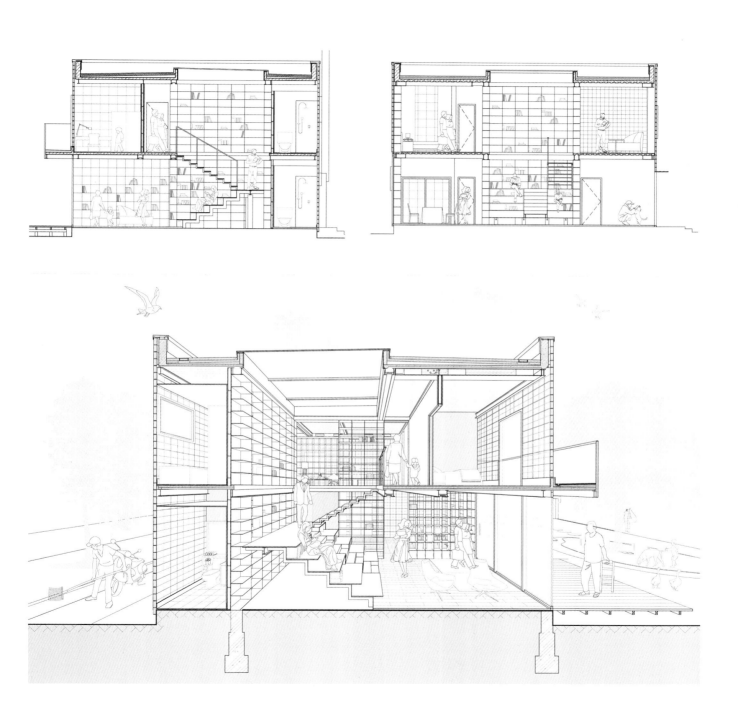

学生：陈铭行，张塑琪，王琪　Student: CHEN Mingxing, ZHANG Suqi, WANG Qi

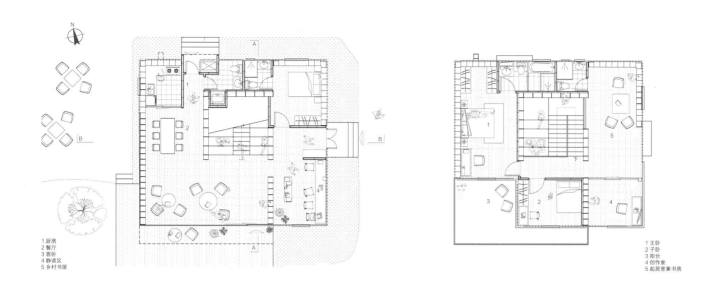

1 厨房
2 餐厅
3 客卧
4 静读区
5 乡村书屋

1 主卧
2 子卧
3 阳台
4 创作室
5 起居室兼书房

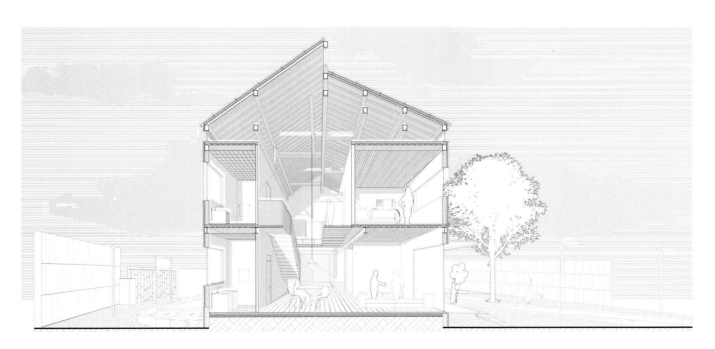

学生：王瑞蓬，翁鸿祎，袁振香　Student: WANG Ruipeng, WENG Hongyi, YUAN Zhenxiang

建筑设计研究（一）ARCHITECTURAL DESIGN RESEARCH 1

基本设计
BASIC DESIGN

金鑫
JIN Xin

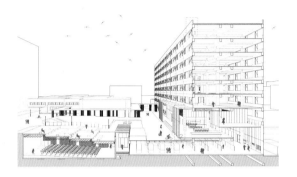

研究问题

建筑系馆是开展建筑教育的"空间容器"，也是一种重要的教学资源。它不但提供了学习、交流场地，也是学生理解和体验空间的最直接之所。随着建筑教育理念的改变，教学空间需求也随之而变化。学习行为不再局限于专业教室，教学场所更加多元开放。南京大学建筑与城市规划学院大学文科楼原本并非专门为建筑学设计，在2014年左右学院对其进行大刀阔斧的改造，使其面貌焕然一新，堪称系馆改造的典范。随着时间的推移，建筑与城市规划学院对使用空间又提出新的要求。面对更新，并非一味推倒重来。如何既解决空间使用问题，又能保留不同时期的历史痕迹，还能创造新的可能，这是此次课程研究的重点。

功能重组

新的建筑学教育理念带来新的空间和形式的需求。学生需要提前了解文科楼在历史的不同时期的空间状态和使用用途，尤其关注构造细节与施工、改造痕迹。从空间功能上而言，教学场所不再单纯表现为从前单一的教室功能，而更加多元开放。教师和学生的行为模式发生变化，对互动交流需求的增加，从而对灵活多变、模式各异的交流空间提出要求。此外还有展示空间、实验室空间、模型制作空间等新功能，都对建筑的水平空间与垂直空间提出新的要求。课程邀请学院老师作为业主，提出需求。学生针对原有建筑空间，新增需求空间，进行选择组合。

形式"外卷"

在改造空间中运用多样的形式语言，能较为直观地反映空间属性，包括结构形式、围护形式、装饰元素等。同时要求在一定程度上保留建筑空间中不同时期的形式，让使用者感知改造的时间性。

研究目的

遗产保护和建筑考古学中有"地层学"这个概念，但在建筑改造设计中，更多的是一种选择性的形式表达。这个课程通过对建筑环境和单体的历史研究，选择性地对不同时期的改造形式进行保留，同时对建筑空间形式进行创新，以此引发学生对于建筑改造设计的深入思考。

Research question

Architecture departmental pavilions are a "spatial container" for the delivery of architectural education and an important teaching resource. It not only provides a venue for learning and communication, but also the most direct place for students to understand and experience space. As the concept of architectural education has changed, so have the needs of the teaching space. The act of learning is no longer confined to specialist classrooms, and teaching spaces are becoming more diverse and open. The University Liberal Arts Building at Nanjing University's School of Architecture and Urban Planning was not originally designed as a teaching space specifically for architecture, but around 2014 the School undertook a radical transformation of the building, giving it a new look that is a model for

departmental buildings. Over time, the School of Architecture and Urban Planning has made new demands on the space it uses. Faced with renewal, it is not just a matter of pushing back. How to solve the problem of using the space while preserving traces of the history of the different periods and also creating new possibilities is what this course is looking at.

Functional reorganisation
The new concept of architectural education brings with it the need for new spaces and forms. Students need to know in advance the spatial state and use of the Liberal Arts Building in different periods of its history, paying particular attention to construction details and traces of construction and renovation. In terms of spatial function, it is no longer simply a single classroom, but a more diverse and open place for teaching and learning. The behavioural patterns of teachers and students have changed and the need for interaction and communication has increased, requiring flexible and varied space for communication. There are also new functions proposed such as display space, laboratory space and model-making space, all of which will place new demands on the horizontal and vertical spaces of the building. The course invites faculty members to act as owners and propose requirements. Students select combinations of new demanded space in response to existing building space.

Formal "out-of-volume"
The use of a variety of formal languages in the renovated space can reflect the properties of the space in a more intuitive way, including structural forms, envelope forms, decorative elements, etc.. At the same time, it is required to retain a certain degree of forms from different periods in the building space to give the user a sense of the temporality of the transformation.

Study purpose
The concept of "stratigraphy" is used in heritage conservation and architectural archaeology, but in the design of building renovations it is more of a selective formal expression. In this course, through the historical study of the architectural environment and single building, selective forms of renovation from different periods are preserved, while innovative forms of architectural space are created, thus provoking students to think deeply about the design of architectural renovation.

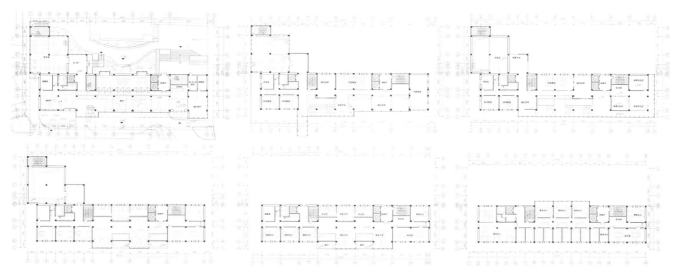

学生：潘晴，张梦冉，王雨嘉　Student: PAN Qing, ZHANG Mengran, WANG Yujia

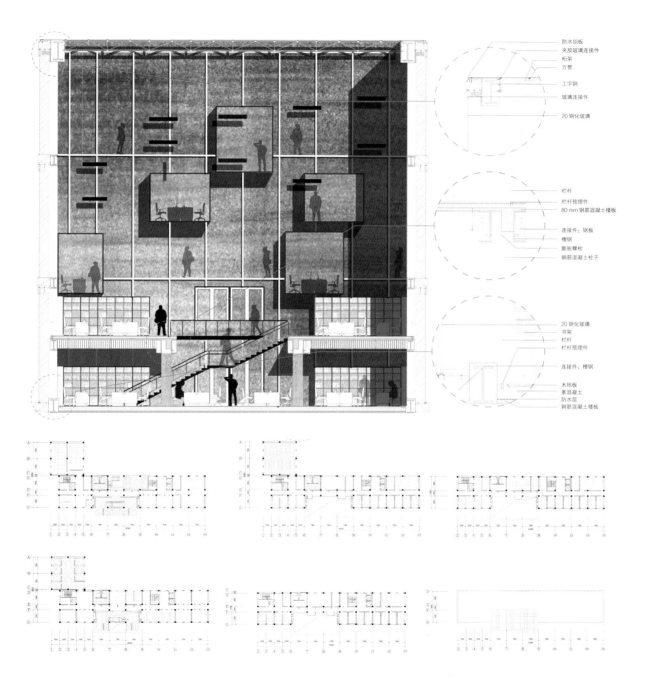

学生：丁明昊，陈茜，任钰佳　Student: DING Minghao, CHEN Qian, REN Yujia

概念设计
CONCEPTUAL DESIGN
鲁安东
LU Andong

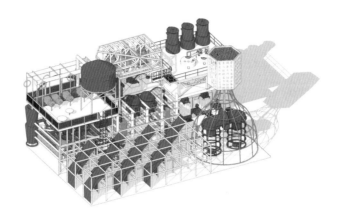

研究问题：增强场所

如何理解当代城市的日常空间？传统意义上的非物质要素正变得可感可触，无处不在的技术正日益成为人的基本能力的外延，并决定了人与世界的基本交互的形式，超物的在地显现使得每一个场所都无法仅作为它自身被认知……这些都构成了场所真实性的一部分，而场所的物质维度则成为或回归于它应有的"中介"本质。现代建筑对于独立存在的物质维度的预设以及在此基础上发展出的大量设计手段和设计价值是否依然有效？我们如何真正地为人设计？我们如何为真正的人设计？

载体对象：文学之都

2019年南京获得联合国教科文组织的"世界文学之都"称号。"世界文学之都"城市空间计划（2020—2023）是在"全球创意城市网络"的国际化语境下，多维度运用文学资源为城市赋能的行动计划。本课题将在此实际项目框架下，探索在地化设计如何一方面为本地赋能，另一方面充分运用增强场所的新行动机遇发挥更大的作用。

课程信息和进度

学生人数：2人一组，自选场地进行设计。
上课时间：每周二上午9—12点；每周三下午2—5点。
教学包括课上及线上"中介"主题系列讲座6次。

第一周 9.15/9.16：课后——增强场所构成分析图解。
第二周 9.22/9.23：汇报增强场所图解及视觉论文；课后——文学之都"赋能"路径。
第三周 9.29/9.30：汇报文学之都"赋能"路径；课后——增强场所初步规划设计。
第四周：国庆放假。
第五周 10.13/10.14：汇报增强场所初步规划设计；10.17：结合圆桌线上中期评图。
第六周 10.20/10.21：汇报根据评委意见修改后的设计；汇报出图计划。
第七周 10.27/10.28：图纸深化。
第八周：终期评图（线下；每组不少于4张A1）。

Research question: enhancing place

How to understand the daily space of the contemporary city? Traditionally immaterial elements are becoming palpable, ubiquitous technology is increasingly becoming an outgrowth of basic human capacities and determining the form of basic human interaction with the world, and the local manifestation of the supra-

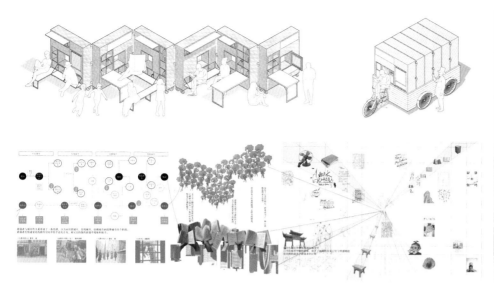

physical makes every place impossible to be perceived only as itself...These form part of the reality of place, and the material dimension of place becomes or returns to its proper "mediating" nature. Is the modern architectural presupposition of a separate material dimension, and the vast array of design tools and design values that have developed on this basis, still valid? How do we truly design for people? How do we design for real people?

The carrier object: the capital of literature
In 2019 Nanjing was awarded the UNESCO title of "World Capital of Literature". The "World Capital of Literature" Urban Space Programme (2020–2023) is an action plan for the multi-dimensional use of literary resources to empower cities in the international context of the "Global Creative Cities Network". Within the framework of this practical project, this topic will explore how localised design can, on the one hand, empower the local context and, on the other hand, make full use of new opportunities for action that enhance place.

Course information and progress
Number of students: two students working in pairs to design a site of their choice.

Class time: Tuesdays 9am–12am; Wednesdays 2pm–5pm. The course consists of a series of 6 in-class and online lectures on the topic of "mediation".

Week 1 9.15–9.16: After class—illustration of enhanced site composition analysis.
Week 2 9.22–9.23: Presentation of enhanced site illustration and visual essay; after class— "Empowerment" pathways of the literary capital.
Week 3 9.29–9.30: Presentation on the "empowerment" pathway of the literary capital; after class—preliminary planning and design of the enhanced site.
Week 4 National Day holiday.
Week 5 10.13–10.14: Presentation on the preliminary planning and design of the enhanced site; 10.17: Mid-term drawing evaluation in conjunction with the round table online.
Week 6 10.20–10.21: Presentation of revised design based on comments from the jury; presentation of the plot plan.
Week 7 10.27–10.28: Deepen the drawings.
Week 8: Final drawing evaluation (offline; no less than 4 A1s per group).

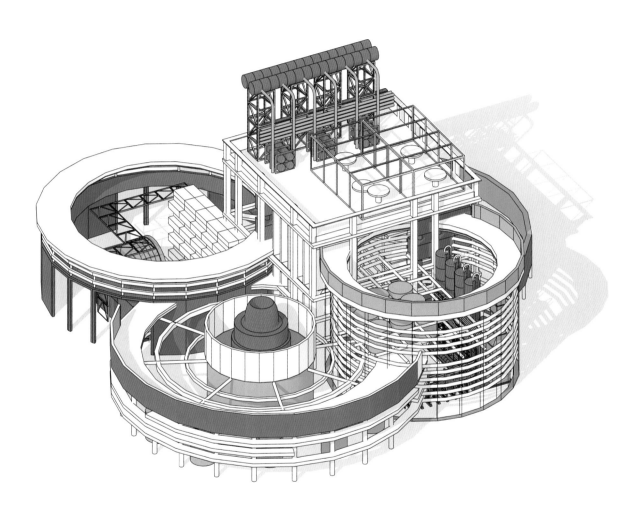

学生：龚豪辉，庞馨怡，王蕾 Student: GONG Haohui, PANG Xinyi, WANG Lei

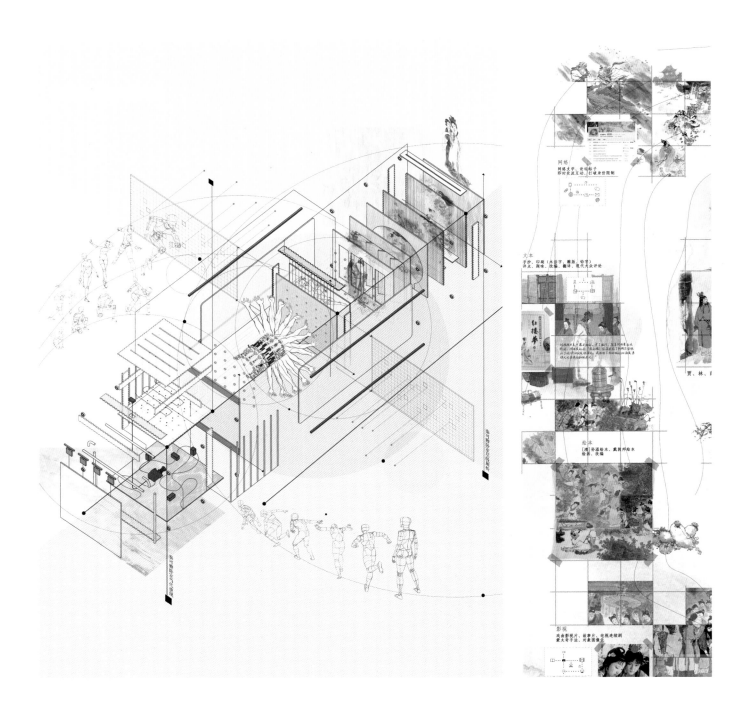

学生：周理洁，唐敏，任钰佳　Student: ZHOU Lijie, TANG Min, REN Yujia

概念设计
CONCEPTUAL DESIGN
周渐佳
ZHOU Jianjia

课程内容

疫情使得人们对线上空间的适应与使用突然加速。线上空间曾经被认为是物理空间的附属，却在很短的时间内成为所有活动发生的重要乃至唯一载体。这个过程也将线上空间推到了建筑学科的面前，我们会发现线上空间是与物理空间并行，且有着同等意义的一个新领域，此前极少获得来自学科的关注。这里所说的线上空间不是超越了电影、游戏中作为背景的场景设计，而是突破物理空间的限制，去架构一种新的空间逻辑和交互手段（例如空间的嵌套、循环，时间的重置等），这些成果同样能在物理空间中制造反差，互为验证。为了回应多年来概念设计课程的积累，本次课程以一场展览作为设计对象，以线上空间作为命题的大背景展开。同学们将在为期8周的时间内围绕线上展览的空间逻辑、交互手段、相关技术等做全方位的讨论。最终成果同样形成展览，在线上与线下空间同时展示。课程中的所有讲座、小论文、讨论、手稿都将汇编成册，作为对课程的记录。

讲座

交互：凯尔，彼真科技创始人，交互技术专家。
数字生存：陆扬／袁松，青年艺术家。
场景：张润泽，TikTok AR Platform。

课程安排

课程以三次评图为节点，划分成三个阶段。在每两次集中评图之间，围绕着一个特定主题，例如概念、工具、技术手段等，以类似 Workshop 的方式进行密集工作。课程采用单人或双人合作的方式，视选课人数而定。

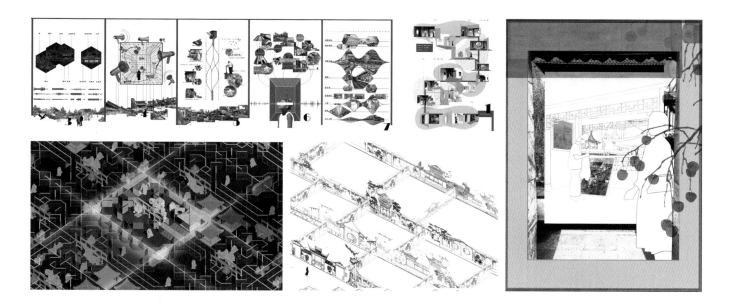

Course content

The epidemic has led to a sudden acceleration in the adaptation and use of online spaces. Online space was once considered an adjunct to physical space, but it has in a very short period of time become an important and even the only carrier for all activities to take place. This process has also brought online space to the forefront of the architectural discipline, and we find that online space is a new field that is parallel to physical space and has equal significance, having received very little attention from the discipline before. The online space referred to here goes beyond the design of scenes as backdrops in films and games, but goes beyond the limits of physical space to structure a new spatial logic and means of interaction (e.g. nesting and looping of spaces, resetting of time, etc.), the results of which can also create contrasts and validate each other in the physical space. In response to the accumulation of conceptual design courses over the years, this course takes an exhibition as the design object, with the online space as the broader context for the proposition. The students will discuss all aspects of the spatial logic, means of interaction and related technologies of the online exhibition over a period of 8 weeks. The final result will also be an exhibition, which will be presented in both the online and offline spaces. All the lectures, short papers, discussions, and manuscripts from the course will be compiled into a book as a record of the course.

Lecture

Interaction: Kyle, founder of Pizen Technology, interaction technology expert.
Digital survival: LU Yang/ YUAN Song, young artist.
Scene: ZHANG Runze, TikTok AR Platform.

Course arrangement

The course is divided into three phases with three drawing evalution as nodes. In between each of the two drawing evalution, intensive work is carried out in a workshop-like manner around a specific theme, such as concepts, tools, technical means, etc.. The course is conducted in single or double students depending on the number of participants.

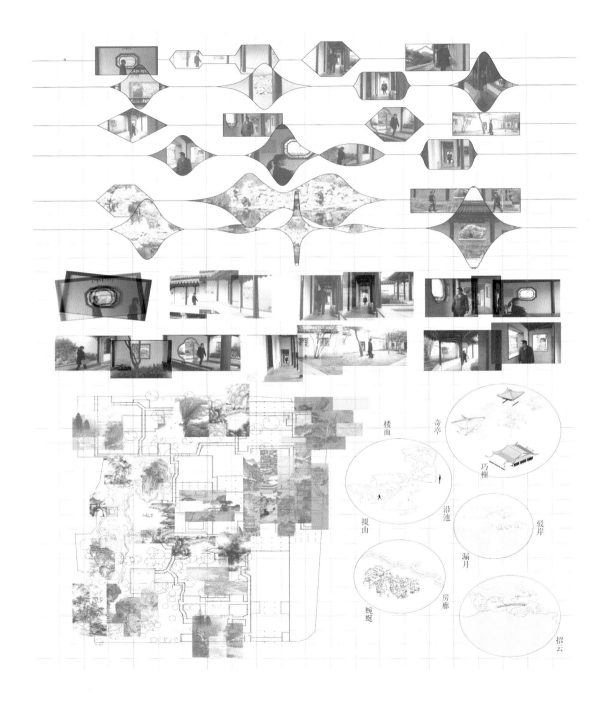

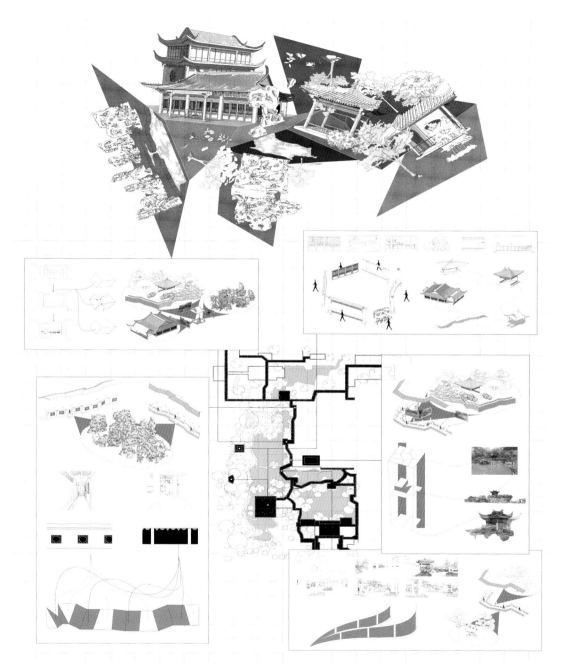

学生：陈茜，刘雪寒 Student: CHEN Qian, LIU Xuehan

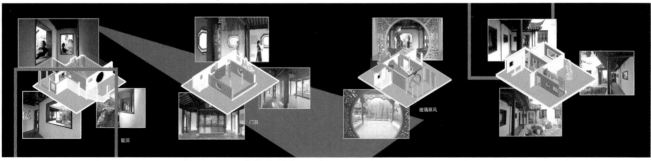

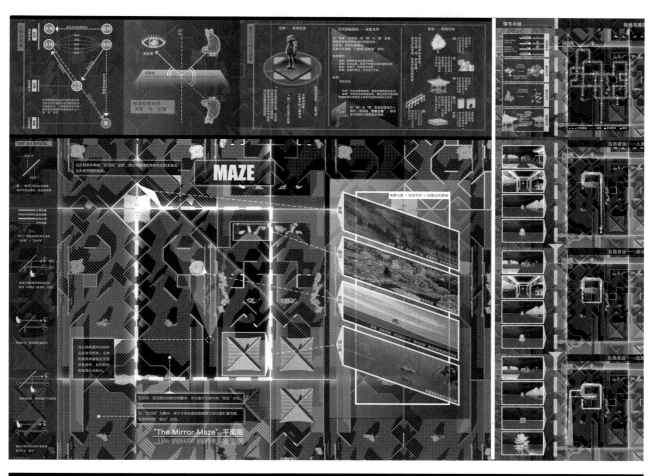
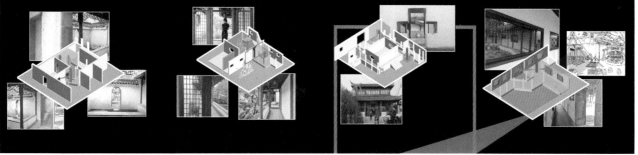

学生：张云松，陈婧秋　Student: ZHANG Yunsong, CHEN Jingqiu

建筑设计研究（二）综合设计 ARCHITECTURAL DESIGN RESEARCH 2 COMPREHENSIVE DESIGN

南京大学鼓楼校区学习中心设计
LEARNING CENTER DESIGN IN GULOU COMPUS OF NANJING UNIVERSITY

程 超
CHENG Chao

教学目标

该课程设计以学生熟悉的校园建筑类型为主题，基于校园文脉传承、现状空间格局和未来学习需求，参照校园发展趋势和相关案例，从场所与活动、空间和功能、视线和路径等关系入手，判别设计解决的核心问题，提出明晰的设计策略，强化准确的专业表达，符合相关规范的要求。设计成果要求达到初步设计深度，并选择适合当下建筑产业发展趋势的技术重点进行专项设计。

设计主题

校园文脉 / 交互网络 / 主动式建筑 / 装配化设计 / 机电专业协同

区位和背景

位于城市中心的南京大学鼓楼校区空间结构独特，被城市道路划分为南北两区并与社区比邻。校园空间经过不同历史阶段的演进，形成双轴共生的格局。靠近校门和城市道路的图书馆既是双轴转换的重要节点，也是校园对外展示交流的窗口。

建于1987年的新图书馆与1937年建成的老图书馆（原中央大学图书馆）以书库进行连接，形成一个以书库为中心向西南两翼发展的布局。2001年图书馆改造设计在扩大阅览空间的同时，将老图书馆改为校史馆，然而以书库为中心的格局并未改变。仙林校区成为主校区后，鼓楼校区图书馆功能弱化，原书库基本停用。基于鼓楼校园新的发展定位，考虑拆除原图书馆及书库部分，整合现有校史馆功能，规划建设新型的校园学习中心。

基于未来校园发展趋势的学习中心，是集合阅览、自修、研讨、社交、餐饮等功能为一体的多功能综合性建筑，以学生的自主学习为中心，促进校园文化的传播和跨学科的交流融合。它将成为面向未来的学习聚落和知识集市，把学习行为扩展为学习社交。新功能的介入将导致对校园空间的重新诠释，从图书馆到学习中心的转化，是从"以书籍为中心"到"以学生为中心"的转化，建筑的核心功能从各类专业书籍的搜集转化为校园空间网络的交集。

教学进度

第一周：基地调研及场地分析（讲解校园空间格局历史的演化以及新老图书馆布局的演变）；
第二周：提出功能策划，比较可能的概念方案；
第三周：概念方案；
第四周：深化设计（讲解三个技术重点内容及相关规范参考）；
第五周：中期考核；
第六周：绘制图纸（讲解总平面图和单体设计对图纸深度的具体要求）；
第七周：设计表达；
第八周：答辩准备。

Teaching objectives

The course design takes the familiar campus building types as the theme, based on the heritage of the campus, the current spatial pattern and the future learning needs, with reference to the campus development trend and relevant cases, starting from the relationships between places and activities, space and functions, sight lines and pathways, etc., to identify the core issues to be addressed by the design, propose clear design strategies, strengthen accurate professional expressions and meet the requirements of relevant codes. The design results are required to meet the preliminary design depth requirements and to select technical priorities for special design that are suitable for the current trends in the construction industry.

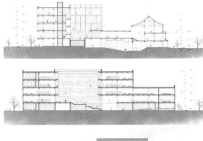

Design themes
Campus context / interaction network / active architecture / assembly design / electromechanical collaboration

Location and context
Located in the heart of the city, the Gulou Campus of Nanjing University has a unique spatial structure, divided by urban roads into two zones to the north and south and adjacent to the community. The campus space has evolved through different stages of history to form a twin-axis symbiosis. The library, close to the campus gate and the city road, is an important node in the transition between the two axes, as well as a window for the campus to showcase and communicate with the outside world.

The new library, built in 1987, is connected to the old library (formerly the library of National Central University) built in 1937 by a bookstore, forming a layout in which the center of the bookstore develops into two wings to the southwest. With the renovation of the library in 2001, which was designed to expand the reading space and change the function of the old library into the university history museum, the pattern of the bookstore as the center still remained unchanged. After the Xianlin Campus became the main campus, the library function of the Gulou Campus was weakened and the former bookstore was basically decommissioned. Based on the new development orientation of the Gulou Campus, consideration was given to demolishing the former library and bookstore part, integrating the function of the existing University history museum and planning the construction of a new campus learning center.

Based on the future development trend of the campus, the learning center will be a multi-functional and comprehensive building integrating reading, self-study, discussion, social intercourse and catering functions, with students' independent learning as the focus, promoting the dissemination of campus culture and interdisciplinary communication and integration. It will become a future-oriented learning communities and knowledge marketplace, extending the act of learning into a learning social. The new functions will lead to a reinterpretation of the campus space, from library to learning center, from "book-centerd" to "student-centerd". The core function of the building is transformed from a collection of specialist books to a connection of the campus spatial network.

Teaching progress
Week 1: Base research and site analysis (explain the historical evolution of campus spatial pattern and the evolution of the layout of new and old libraries).
Week 2: Propose functional planning and compare possible conceptual plans.
Week 3: Concept plan.
Week 4: In-depth design (explain the three key technical elements and related specification references).
Week 5: Mid-term assessment.
Week 6: Drawing (explain the specific requirements of drawing depth for general layout and individual design).
Week 7: Design presentation.
Week 8: Preparation for the defense.

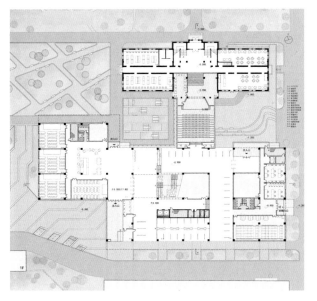
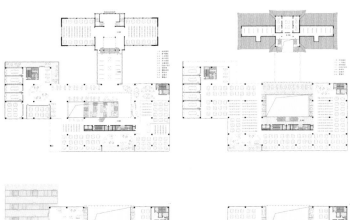

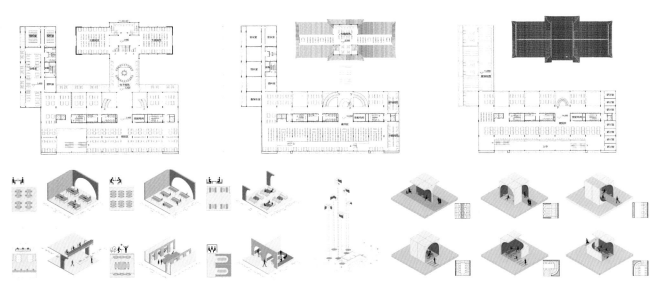

学生：白珂嘉，陈铭行　Student: BAI Kejia, CHEN Mingxing

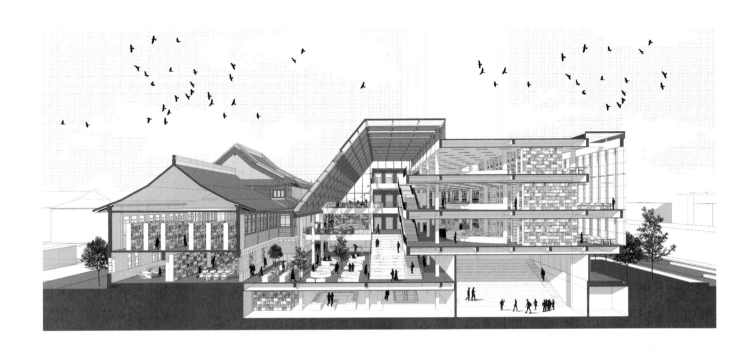
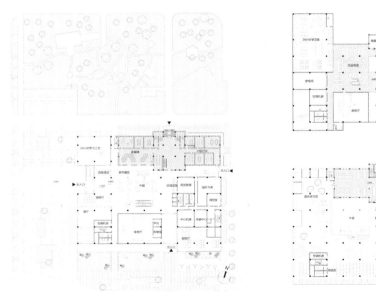

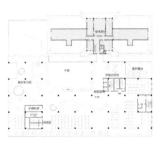

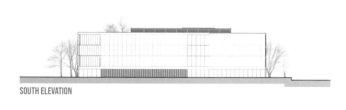

SOUTH ELEVATION

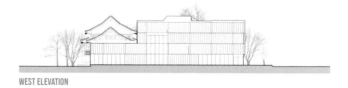

WEST ELEVATION

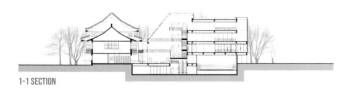

1-1 SECTION

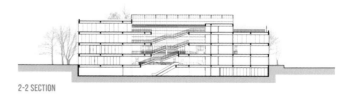

2-2 SECTION

学生:龚豪辉,丘雨辰 Student: GONG Haohui, QIU Yuchen

建筑设计研究（二）城市设计 ARCHITECTURAL DESIGN RESEARCH 2 URBAN DESIGN
宁芜铁路秦虹段改造更新设计研究
STUDY ON RECONSTRUCTION AND RENEWAL DESIGN OF QINHONG SECTION OF NINGWU RAILWAY

华晓宁
HUA Xiaoning

教学目标

基础设施是当代城市研究与实践的重要主题。作为社会生产和为居民生活提供公共服务的物质工程设施及其系统，它保障着城市有机体的运行，同时又是城市物质空间系统的重要组成部分，自身便占据了场址，界定了空间，形成了场所，连接成系统，构筑了场域。以往基础设施被仅仅被视作市政工程的专业领域，遵循工具理性，且被传统建筑学忽视多年，许多基础设施已成为城市中消极和被动的要素。然而对于未来都市而言，它已不再仅仅是边缘化、辅助性和服务性的角色，而是一种重要的城市操作性对象与媒介。如何对其"赋能"，将其转化为城市中更为积极、能动的要素，成为激发城市生活的新型"触媒"，是本课题的主要目标。

对象与场址

宁芜铁路始建于1933年，时称江南铁路。在近90年的时光中，业已深深嵌入南京的城市肌理中。随着时代的发展，在许多城市场址，宁芜铁路对于城市空间造成了割裂，给市民生活带来了负面影响。近年来，宁芜铁路的外迁已被提上议事日程，这为其原有沿线地段的城市空间与市民生活带来了新的机遇。秦虹段是目前现存宁芜铁路与城市居住社区关系最为密切、最为典型的区段，对于探讨城市基础设施重定义为激发城市活力和公共性的触媒这一主题，具有较强的样本价值。

成果要求

对宁芜铁路秦虹段（东起中和桥道口，西至应天大街高架）及其沿线周边环境进行深入调研，分析存在问题与矛盾，了解沿线居民需求，构想该区段未来愿景，自行拟定任务书，提出改造更新策略，完成方案设计。方案须综合考虑城市、建筑、环境景观，进行整合设计。

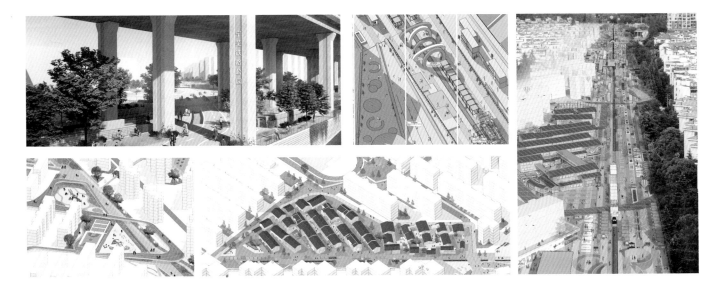

Teaching objectives

Infrastructure is an important theme in contemporary urban research and practice. As a physical engineering facility and system that provides public services for social production and residential life, it guarantees the functioning of the urban organism and is an important part of the physical-spatial system of the city, occupying the site, defining the space, forming the place, connecting the system and constructing the field. In the past, infrastructure was regarded as a specialised field of civil engineering, following an instrumental rationale, and ignored by traditional architecture for many years. Much infrastructure has become a passive and reactive element of the city. For the city of the future, however, it is no longer just a marginal, auxiliary and service role, but an important urban operable object and medium. The main objective of this project is to "empower" it and transform it into a more active and dynamic element of the city, a new type of "catalyst" that will stimulate urban life.

Object and site

The Ningwu Railway was built in 1933, when it was known as the Jiangnan Railway. Over the past 90 years, it has become deeply embedded in the urban fabric of Nanjing. With the development of the times, the Ningwu Railway has caused a fragmentation of the urban space in many urban sites, which has had a negative impact on the life of the citizens. In recent years, the relocation of the Ningwu Railway has been on the agenda, bringing new opportunities for urban space and civic life along its original route. The Qinhong section is the most closely related and typical section of the existing Ningwu Railway to the urban residential community, and has a strong sample value for exploring the theme of redefining urban infrastructure as a catalyst for urban vitality and public character.

Achievment requirements

To conduct an in-depth study of the Qinhong section of the Ningwu Railway (from the Zhonghe Bridge crossing in the east to the Yingtian Street elevated in the west) and its surroundings, analyse the problems and contradictions, understand the needs of the residents along the line, conceptualise the future vision of the section, draw up the mission statement, propose a renovation and regeneration strategy and complete the design. The scheme will be designed in an integrated manner, taking into account the urban, architectural and environmental landscape.

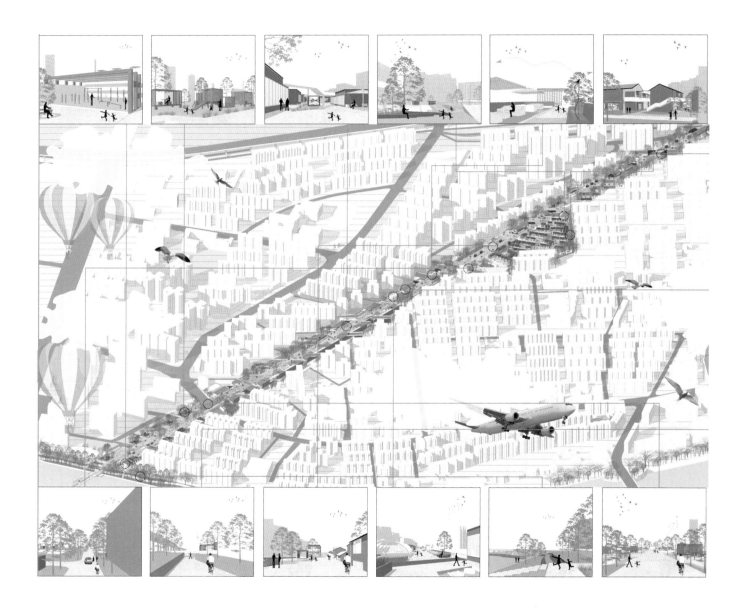

学生：黄翊婕，费元丽，仇佳豪　Student: HUANG Yijie, FEI Yuanli, QIU Jiahao

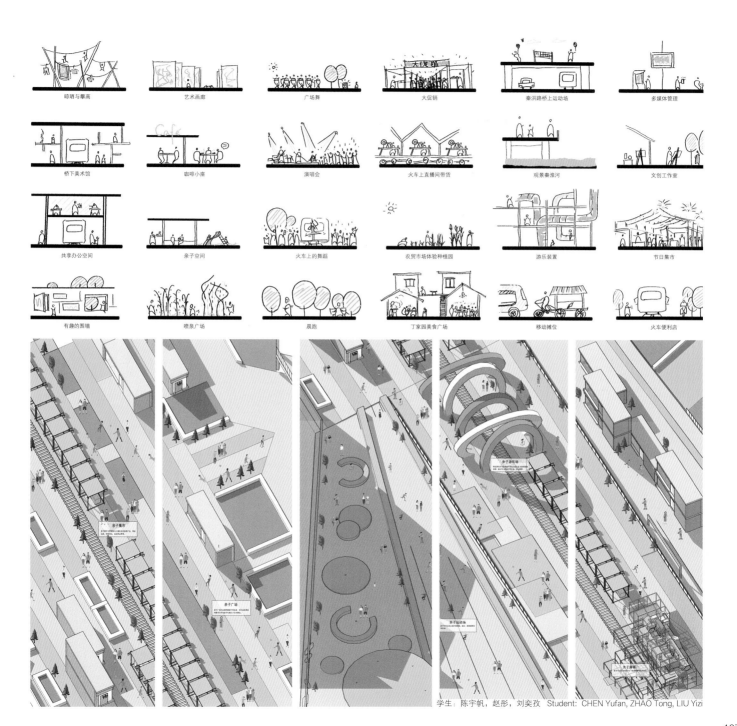

半城市化及其建筑学
50% URBANIZATION AND ITS ARCHITECTURE

胡友培
HU Youpei

背景与问题

城市化以压倒性的优势成为人类几乎唯一的生活方式，它将大片的自然地貌、农地山林圈入城市领地，将城市铺满整个都市区。它以理性而无情的机制以及现代主义的草纸策略，将原本丰富多样、起伏不平的原始地表快速清零，转化为一块块标准的城市土地，从而完成自然物到商品的属性变化。紧跟其后的则是快速生产的一片片无差异的广谱城市，与零度建筑学的填空游戏。这是一种均质化的空间的生产机制。

半城市化是对既有城市化现实的一种悬置与抵抗，是对另一种人居环境可能方式的想象，对低影响度城市化模式的一种探索。它使得在不可阻挡的城市扩张中，自然生态与人工环境间有共存的可能与机会。它是在城市演化到都市区阶段，在中尺度下，关于都市区形态学的一种大胆假想。都市区除了呈现由农田、厂房、园区、新城混合而成的面目模糊的景观，还可能呈现什么样的面目和形式？还可以用什么样的形态同时组织人工物与自然？

任务与限定

半城市化是口号与立场，也是任务与限定。为了给设计试验一个刚性的约束，人为将设计范围内可建设用地缩减为原来的50%，而另一半则保持原本的自然属性。用50%的土地实现100%的人口和建设量，并不是简单地将建筑高度翻番，继续保持自然与城市的分立，而是强迫设计者丢掉所有的城市既有模式与成熟的建筑类型学，在设计的底层，展开艰难的创新。50%并不是教条与精确的计算，而更像是一种象征意义上的武断，为无边的想象设定一个可供锚固的起点。

工作与内容

半城市化要求在两个层面对都市区形态学做出全新的尝试。其一是都市区的架构层次，即突破传统的格网新城模式，构想一种新颖的区域空间组织模式，以更加有效地实现自然系统与人工系统的共存，保证都市生活有序展开。其二是与构架匹配的建筑学层次，即突破传统的城市建筑类型，创造、拼贴、裁剪、再加工各种建筑的和人工物的可能空间形式，以从城市物质形态的底层，为都市区架构提供物质性支撑，并激发可能的半城市化的生活形态。

Background and issues

Overwhelmingly, urbanization becomes almost the only way of human life, it will be a large area of natural landscape, farmland and forest into urban territory, and spreads the city over the entire metropolitan area. With rational and ruthless mechanism and the modernist straw paper strategy, the original rich and diverse, undulating original land is rapidly cleared and transformed into standard urban land, thus completing the change of nature to commodity. This is followed by the rapid production of the Generic City, a broad spectrum of undifferentiated cities, and the zero-degree architecture blank filling game. This is a homogenized spatial production mechanism. Semi-urbanization is a kind of suspension and resistance to the existing urbanization reality, an imagination of another possible way of human settlement environment, and an exploration of low-impact urbanization mode. In order to make the inevitable urban expansion, there is the possibility and opportunity of coexistence between natural ecology and artificial environment. It is a bold assumption about the morphology of the metropolitan area at the mesoscale at the stage of urban evolution to metropolitan area. What face and form could the metropolitan area possibly take on other than a faceless landscape mixed with farmland, factories, parks and new towns? In what form is it possible to organize both artificial and nature?

Tasks and limitations

Semi-urbanization is the slogan and position, and also the task and limitation. In order to provide a rigid constraint for the design experiment, the constructible land within the design scope is reduced to 50% of the original, while the other half is kept with the original natural properties. To achieve 100% population and construction volume with 50% land is not to simply the double the height of buildings and keep the separation between nature and city, but to force designers to discard all existing urban patterns and mature architectural typologies and carry out difficult innovations at the bottom of the design. Fifty percent is not dogma or precise calculation, but rather symbolic arbitrariness, setting an anchor for the boundless imagination.

Work and contents

Semi-urbanization requires a new attempt to the morphology of metropolitan areas at two levels. One is the structure level of metropolitan area. It breaks through the traditional grid new city mode, conceive a new regional spatial organization mode, in order to realize the coexistence of natural system and artificial system more effectively, and ensure the orderly development of urban life. The other one is the architectural level matching with the architecture. It breaks through the traditional urban architectural types, creates, collages, cuts and reprocesses the possible spatial forms of various architectural and artificial objects, so as to provide material support for the metropolitan architecture from the bottom of the urban material form and stimulate the possible semi-urban living form.

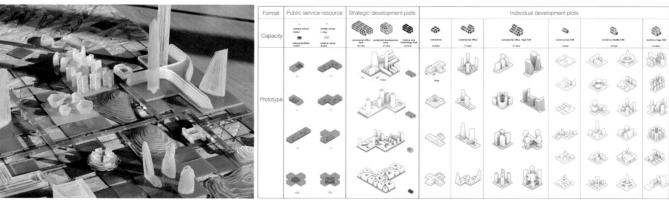

学生：宋贻泽，马丹艺，于智超，赖泽贤 Student: SONG Yize, MA Danyi, YU Zhichao, LAI Zexian

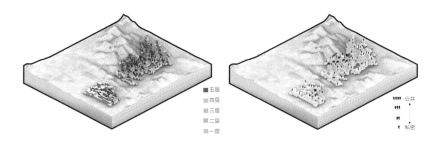
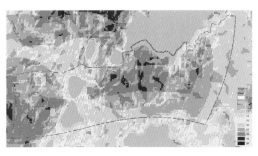

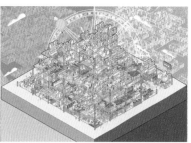

学生： 徐佳楠 方艺璇 赵琳芝 朱激清　Student:Xu Jianan Fang Yixuan Zhao Linzhi Zhu Jiqing

An Urban Sharawaggi 城市设计研究
STUDY ON URBAN DESIGN OF AN URBAN SHARAWAGGI

安东尼奥·彼得罗·拉蒂尼　胡友培
Antonio Pietro Latini, HU Youpei

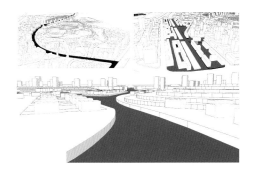

教学目标

工作坊试图探索如何使普通人的生活场所比现在更加愉悦和可持续，重点是环境形式和美学品质。为了克服当代城市令人压抑的单调性和混乱状况，我们依靠中国传统的园林提供的刺激：一种在和谐多样的环境中结合多种环境设计元素的艺术。本次课程主要涵盖了诸如街道/街区格局、图底关系、道路和开放空间类型设计、地块、建筑类型、建筑密度、流线设计等设计主题。

教学内容

课程前期学生能够积累基本的城市设计学科背景知识，且每周进行专题设计。然后在此基础上，由三人一组合作进行城市设计。选址位于南京江宁区东流村，基地面积约 300 hm²。前期的图形练习在最后的小组设计项目中达到高潮，在这个过程中把空间构成、景观设计和规划设计在创新的美学研究中结合起来。

Teaching objectives

The workshop tries to explore how to make ordinary people's living place more pleasant and sustainable than now, focusing on the environmental form and aesthetic quality. In order to overcome the depressing monotony and chaos of contemporary cities, we rely on the stimulation provided by Chinese traditional gardens: an art that combines a variety of environmental design elements in a harmonious and diverse environment. This course mainly covers design topics such as the street / block pattern, map and background relationship, road and open space type design, plot, building type, building density, streamline design and so on.

Teaching content

In the early stage of the course, students can accumulate basic background knowledge of urban design and carry out special design every week. Then, on this basis, three people work together to carry out urban design. The site is located in Dongliu Village, Jiangning District, Nanjing, covering an area of about 300 hm². The early graphic practice reaches a climax in the final group design project. In this process, the space composition, landscape design and planning design are combined in the innovative aesthetic research.

教学过程 Teaching progress

任务一：来自家人或某人的卡片 Task 1: A card from home or "a person's place"

该项训练要求学生提供一张双面 A4 卡片，呈现一个自己熟悉的对自己有一种特殊的归属感的场所。卡片正面可以用各种形式的图像呈现场所的位置，背面用文字表达该场所的意义。

The training requires students to provide a double-sided A4 card to present a familiar place which has a special sense of belonging to themselves. The front of the card can present the location of the place with various forms of images, and the back can express the meaning of the place with words.

任务二：莫尔瓦尼亚村的草图设计 Task 2: Sketch design for a village in Molvanîa

该项训练是在限定条件下进行村庄规划，该村将包括 250 栋住宅楼，容纳多达 1 000 人，学生利用 250 个 12 mm×6 mm×6 mm 的木块尝试不同的村庄排布方式，其中包括教堂、超市、学校、活动中心 4 栋公共建筑，并规划主要街道、广场和城市花园。

The training is to carry out village planning under limited conditions. The village will include 250 residential buildings and accommodate up to 1 000 people. Students use 250 12 mm×6 mm×6 mm wood blocks to try different village layout methods, including four public buildings of church, supermarket, school and activity center, and plan main streets, squares and urban gardens.

任务三：城市组成的新兵训练营 Task 3: Urban components boot camp

该项训练旨在理解关于城市组成部分和相关指标的一些基本知识——图底关系、街道类型、街区类型，选取了南京市来凤多层小区、黄山路高层小区、中华门混合住区、老门东历史街区四个代表性住区。学生进行现场调研测绘，感受城市的尺度关系。

The training aims to understand some basic knowledge about urban components and related indicators—figure and ground relationship, street type, block type. Four representative residential areas of Laifeng multi-storey community, Huangshan Road high-rise community, Zhonghuamen mixed residential area and Laomendong East historical block in Nanjing are selected. Students conduct on-site survey and mapping to personally experience the scale relationship of the city.

任务四：移除艺术的城市设计 Task 4: Urban design as the art of removing

该项训练的目标是通过"移除"来设计城市的一部分。我们把城市想象成一个坚固连续的体量，其中的空间必须被雕刻出来，雕刻操作后留下的是建筑系统。因此，最开始设计的其实是一个城市的公共空间系统。学生需要制作一个 1:2000 的代表面积 600 m×600 m 理想城市区域的模型。该模型包括一个内部尺寸为 30 cm×30 cm 的正方形框架，深度为 3 mm，内部填充黏土。从黏土层中雕刻出城市空间，雕刻时移除的黏土被用来建造建筑物。

The goal of this training is to design a part of the city by "removing". We imagine the city as a solid and continuous volume, in which the space must be removed, and the architectural system is left after the carving operation. Therefore, the initial design is actually a city's public space system. Students need to make a 1:2000 model representing an ideal urban area of 600 m×600 m. The model includes an internal size of 30 cm×30 cm square frame, 3 mm deep, filled with clay. The urban space is carved from the clay layer, and the clay removed during carving is used to build buildings.

任务五：南京紫东：如何获得城市园林艺术 Task 5: Nanjing-Zidong: How can we achieve an urban Sharawaggi?

最终的设计是基于"urban Sharawaggi"研究主题的城市局部区域设计。Sharawaggi 的理念和中国经典环境设计的基本构图原则是本次城市设计的出发点。选址位于南京紫东区，在市中心的外围，距离南京中央商务区以东 15 km。场地边界北至仙林大道，西至绕城高速，南至速蓉公路，东至七象河。目前，该场地分布有稀疏的居住区和工业区。地形复杂，是南京典型的丘陵地区。设计内容为混合用途社区，包括各种类型的住宅建筑，还必须包括公共空间（步行街道网络、街区绿地和停车场）以及不同用途的公共建筑和配套服施务。

The final design is the urban local area design based on the research theme of "urban Sharawaggi". Sharawaggi's concept and the basic composition principles of Chinese classic environmental design are the starting point of this urban design. The site is located in Zidong District, Nanjing, on the periphery of the city center, 15 km east of Nanjing central business district. The site borders Xianlin Avenue in the north, ring expressway in the west, Surong Highway in the South and Qixiang River in the East. At present, the site is distributed with sparse residential areas and industrial areas. The terrain is complex, which is a typical hilly area in Nanjing. The design content is mixed-use community, including various types of residential buildings. The block must also include public space (pedestrian street network, block green space and parking lot) and public buildings and supporting services for different purposes.

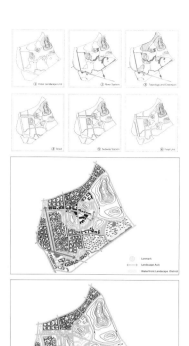

学生：吕广彤，曾敬淇，马致远　Student：Lü Guangtong　ZENG Jingqi　MA Zhiyuan

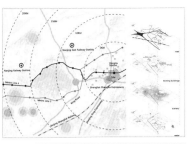

学生：朱激清，王琪，于文爽　Student：ZHU Jiqing, WANG Qi, YU Wenshuang

研究生国际教学工作坊 POSTGRADUATE INTERNATIONAL DESIGN STUDIO

电影建筑工作坊
CINEMATIC ARCHITECTURE WORKSHOP

弗朗索瓦·潘兹　鲁安东
Francois Penz, LU Andong

受 2021 年度南京大学国际合作与交流处"引进人文社科类资深海外专家重点支持计划"支持，剑桥大学弗朗索瓦·潘兹教授开设线上研究生课程"电影建筑"，本专业选课学生 26 人。

课程包括学术讲座系列和影像拍摄工作坊两个版块：学术讲座系列共 8 讲，包括 5 次内部讲座及 3 次线上直播讲座，听课总人数超过 9 600 人次；影像拍摄工作坊进一步扩大参加范围，面向南京大学校内其他学科进行了招募。工作坊时长 7 d，由学生 4 人一组，以南京为研究对象，制作 90 s 短片呈现了后疫情时代公共空间的四个维度。拍摄成果进行了线上评图，参加人数超过 20 000 人。此外，成果短片受邀在 2021 年第 27 届世界建筑师大会放映，受到主办方的高度赞誉。

主讲人：弗朗索瓦·潘兹教授

弗朗索瓦·潘兹是剑桥大学建筑系终身教授、动态影像可视化领域开创者。曾任剑桥大学建筑系主任，剑桥大学马丁研究中心主任，剑桥大学设计、可视化与交流数字实验室主任。他曾主持英国艺术与人文研究基金重大课题"空间文化差异的电影想象"、欧盟第六框架重大课题"新千年的新媒体"等开创性的重大科研计划。2009 年荣获"法国学术界棕榈叶勋章"，该勋章是法国学术界、教育界的最高荣誉。

Supported by the "Key Support Program for the Introduction of Senior Overseas Experts in Humanities and Social Sciences", a key university-level intellectual attraction project in 2021, Professor Francois Penz from the University of Cambridge offered the online graduate course "Cinematic Architecture", which was attended by 26 students.

The course consists of two sections: academic lecture series and video shooting workshop. The academic lecture series consists of eight lectures, including five internal lectures and three online live lectures, with a total attendance of more than 9 600; the video shooting workshop further expanded the scope of participation and was open to other disciplines within Nanjing University. The workshop lasted 7 days and was conducted in groups of 4 students, with Nanjing as the research object. Taking Nanjing as the research ovject, a 90s video was produced to present four dimensions of public space in the post-epidemic era. The results of the filming were evaluated online with over 20 000 participants. In addition, the resulting short film was invited to be screened at the 27th World Congress of Architects in 2021 and received high praise from the organizers.

Speaker: Prof. Francois Penz

Francois Penz is a tenured Professor of Architecture at the University of Cambridge and a pioneer in the field of moving image visualization. He was formerly the head of the Department of Architecture at the University of Cambridge, the Director of the Martin Research Center at the University of Cambridge, and the director of the Digital Laboratory for Design, Visualization and Communication at the University of Cambridge. He once led major research projects such as a Cinematic Musée Imaginary of Spatial Cultural Differences, an Arts and Humanities Research Fund project, and New Media for a New Millennium, an EU Sixth Framework project. In 2009, he was awarded the "Order of the Palm Leaf", the highest honor in the French academic and educational world.

学生:潘晴,马丹艺,邢雨辰 Student: PAN Qing, MA Danyi, XING Yuchen

学生：王锴，王赛施，李心仪　Student: WANG Kai, WANG Saishi, LI Xinyi

研究生国际教学工作坊 Postgraduate International DESIGN STUDIO

建构共生与未来环境建造——国际之声工作坊
CONSTRUCTING COEXISTENCE AND FUTURES OF ENVIRONMENT-MAKING—INTERNATIONAL VOICES WORKSHOP

凯瑞·希瑞斯　丁沃沃
Cary Siress, DING Wowo

课程概述　Course overview

　　2020 年对于世界各国来说是极为不平凡的一年，蔓延到全世界且至今尚未结束的疫情证实了人们在世纪之初的预感，世界正在巨变。此时，南京大学建筑学迎来了办学 20 周年。20 年的历程不长，但充满了探索，面对巨变的世界，南大建筑将肩负责任，重新踏上新的探索征程。建筑学在探索，建筑学需要探索，建筑学需要在探索中重构。这是一个全新的征途，目标和行动都充满不确定性和挑战，也充满期待。为此，我们搭建了国际前沿系列讲座的平台，邀请国际一流大学中对此有思考的学者，尤其是有思想的年轻的学者，请其针对人类共同的生存环境的问题阐述自己的观点，在论述中放出思想的火花。

The year 2020 is an extraordinary year for all countries in the world. The epidemic that has spread all over the world and has not yet ended has confirmed people's hunch at the beginning of the century that the world is changing dramatically. At this time, Nanjing University Architecture ushered in its 20th anniversary. The 20-year course is not long, but it is full of exploration. In the face of a world of great changes, NJU architecture will shoulder the responsibility and embark on a new journey of exploration. Architecture is exploring, architecture needs to be explored, and architecture needs to be reconstructed in exploration. This is a new journey, with goals and actions full of uncertainties, challenges and expectations. To this end, we have set up a platform for a series of international cutting-edge lectures, and invited thoughtful scholars from world-class universities, especially thoughtful young scholars, to elaborate their views on the problems of the common living environment of mankind, and to spark their thoughts in the discussion.

建构共生与未来环境建造：国际之声 Constructing co-existence and futures of environment-making: international voices

　　本系列讲座由 6 场独立演讲和两个圆桌论坛组成，每周三晚 21:00—22:00，共持续 8 周。讲座将于 2021 年 3 月 3 日—4 月 21 日期间在 Umeet 会议平台线上进行。

The lecture series consist of six separate lectures and two round-table forums, delivered every Wednesday from 21:00 to 22:00 (China GMT +8) on Wednesday for 8 weeks. The lecture will be held online on the Umeet conference platform from March 3rd to April 21st 2021.

讲座介绍 Lecture introduction

　　工业革命后，城市的规模和数量开始增长，城市与腹地的关系发生了巨大变化，先是在欧洲，接着是在美洲，后来是亚洲和非洲。城市研究学者、社会学家、地理学家、生态学家和气候学家们指出：20 世纪和 21 世纪城市产生的资源需求超越了它们各自的腹地、区域和生物群所能提供的极限，正威胁着地球的边界。由此产生的聚落模式，在规模、范围和程度上都是前所未有的，同时，也预示着全新的城市化形式。

　　目前对城市化的规范性观点往往以城市为中心，限制了人们观察、分析和积极参与这些新形式的能力。本场讲座将概述以城市为中心对城市化理解的削弱效应，并探讨能够更好地理解城市与腹地间关系的其他城市化途径的可能性。

演讲人：斯蒂芬·凯恩斯　Speaker: Stephen Cairns

After the industrial revolution, the size and number of cities began to grow, and the relationship between cities and hinterland changed dramatically, first in Europe, then in America, and then in Asia and Africa. Urban researchers, sociologists, geographers, ecologists and climatologists point out that the resource demand generated by cities in the 20th and 21st centuries has exceeded the limits that their hinterland, regions and biota can provide, and is threatening the boundaries of the earth. The resulting settlement mode is unprecedented in scale, scope and degree. At the same time, it also indicates a new form of urbanization.

At present, the normative view of urbanization is often city centered, which limits people's ability to observe, analyze and actively participate in these new forms. This lecture will outline the weakening effect of city centered understanding of urbanization and explore the possibility of other urbanization approaches that can better understand the relationship between cities and hinterland.

Agropolitan Territories, Bioregions, and other Settlement Systems: Urbanization After City-Centricity

农业都市领域、生物区域和其他住区系统：以城市为中心后的城市化

IVLS

讲座介绍　Lecture introduction

全球城市化、全球市场一体化和地球生态系统都陷入了一种不断恶化的共生关系中，这种共生关系既不确定，也不稳定。人类精心策划的环境以其多样的形式和规模引发了很大程度的分裂、分离与杂交，以致于混淆了我们对地球不稳定状况的反应。然而，对于许多现代设计梦想家来说，人类世即将发生的生态灾难的叙述方兴未艾，这使得一个星球因为处于危机之中，可以而且必须通过强大的地理建筑调解来重塑。

本场讲座将探讨在将发展重心转移到一个整体的、全方位的行星重建项目中"总体规划复合体"的意义。这个项目以整个地球作为高科技设计知识、管理规划干预和机会主义空间治理的对象。

演讲人：卡里·萨利斯　Speaker: Cary Siress

Planet-wide urbanization, the globally integrated market, and Earth's ecosystems are locked in deteriorating symbioses that are indeterminate and far from stable. Human-orchestrated environment making in its varied forms and scales has provoked such a degree of fragmentation, dissociation, and hybridization as to confound any coherent response to the precarity of our planetary condition. Yet, for many a modern-day design visionary, the ascendant anthropocene narrative of impending ecological catastrophe has rendered a planet that, because it is in crisis, can and must be remade through the prowess of geo-architectural mediation.

This lecture will discuss the significance of the "masterplan complex" in shifting the development focus to an overall and all-round planetary reconstruction project. This project takes the whole earth as the object of high-tech design knowledge, management planning intervention and opportunistic space governance.

讲座介绍　Lecture introduction

这部反映人类世界和我们居住的星球的电影是该系列的第三讲。非语言的电影片段探索了地球共存的平凡而不可思议的一面，展示了我们在时间和空间上的共同经历在多大程度上是令人安慰的和毁灭性的。

视觉演讲是我们对自己（自我形象）、对更大的人类社会（社会形象）以及世界本身（世界观）所保持的相互形成关系的"镜子"，在这个世界里，我们作为地球上的利益相关者相互作用。地球和我们在这里的生活的情景，是对我们所创造的这个试探性的和脆弱的世界的批判性反思，无论是在它的"概念"上，还是在它今天的"具体"表现上，也无论好坏。

演讲人：卡里·萨利斯　Speaker: Cary Siress

This filmic reflection on the human-made world and the planet we inhabit constitutes the third lecture of the series. The non-verbal, cinematic passages explore the mundane and miraculous aspects of earthbound co-existence, showing to what extent our shared experiences in time and space can be as consoling and as they are devastating.

The visual lecture serves as a "mirror" of the mutually formative relationships we maintain to ourselves (self-image), to the greater society of humanity (societal image), as well as to the world itself (worldview) in which we all interact as terrestrial stakeholders. The scenes of earth and our life here are critical reflections on the tentative and fragile world we have made, both in its "conception" as well as in its "concrete" manifestation of what it is today, for better or worse.

讲座介绍 Lecture introduction

尽管生态问题日益困扰着当前的城市生产模式，但世界各地的大多数生计仍然受制于发展和经济增长的普遍范式。本讲座将以菲力克斯·加达里的"生态哲学"概念为基础，探讨他的三个生态范畴——环境、社会关系和人类主体性，以推测生态变化的整体愿景。本讲座认为，只有积极参与超越当前建筑和城市规划中资源效率范式的反叙事，同时更多地考虑可能引发社会变化的紧急空间安排，才能实现真正可持续的生态转变。这些"其他"叙事通过鼓励能力建设、合作、亲属关系和关心来动员微观政治变革。从大规模住房到非正式城市发展，我们在世界范围内寻找各种环境制造的例子，追踪环境、社会组织和主体性之间的联系，以解决我们如何动员反叙事，以实现可持续城市生态的必要改变以及如何将这些叙事运用到建筑和城市设计实践中。

演讲人：雷纳·赫尔　Speaker: Rainer Hehl

While ecological concerns increasingly confound current modes of urban production, the majority of livelihoods worldwide are still bound to prevalent paradigms of development and economic growth. Drawing on Felix Guattari's concept of "ecosophy", this lecture will investigate his three ecological registers—the environment, social relations, and human subjectivity, in order to speculate on an integral vision of ecological change.This lecture argues that a shift toward truly sustainable ecologies can only be achieved by actively engaging in counter-narratives that go beyond current paradigms of resource efficiency in architecture and urban planning, while taking more account of emergent spatial arrangements that can trigger social change. These "other" narratives mobilize micro-political transformations by encouraging capacity building, co-operation, kinship, and care. Looking at various examples of environmental making worldwide, from mass housing to informal urban development, connections are traced between environment, social organization, and subjectivity in order to address how we can mobilize counter-narratives for necessary change toward sustainable urban ecologies, and how these narratives can be deployed in architectural and urban design practice.

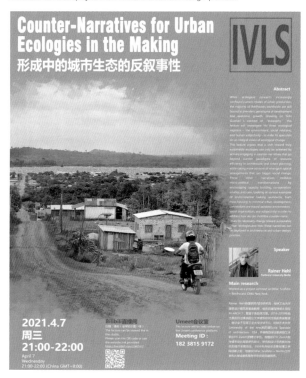

讲座介绍　Lecture introduction

在全球范围内，建筑施工和运营贡献了碳排放的39%，在原材料开采和固体废物生产方面也超过了50%。这使得建筑业成为能源和资源的最大消耗者，也是排放和废物的最大制造者。为了克服线性经济系统中潜在的社会、经济和环境问题，循环经济的概念越来越引起人们的注意，它被定义为"旨在保持产品、部件及材料在任何时候都具有最高效用和价值，并通过设计来进行的恢复和再生"。

循环经济要求我们在设计、建造和运营建筑的方式上以及在建成环境中管理资源的方式上做出范式的转变。本场讲座将重点介绍康奈尔大学循环建造实验室正在进行的研究项目以及来自德国2hs建筑师和工程师事务所的建筑类型原形，试图引发对当今产业在循环经济中向循环建筑做出必要性和系统性变革的深入讨论。

演讲人：菲力克斯·海斯尔　Speaker: Felix Heisel

Globally, building construction and operation accounts for 39% of carbon emissions, as well as over 50% of raw material extraction and solid waste production, thereby making the construction industry the biggest consumer of energy and resources, as well as the biggest producer of emissions and waste. As a means for overcoming the social, economic, and environmental problems of the underlying linear economic system, the concept of the circular economy is increasingly gaining attention, defined as an approach that is "restorative and regenerative by design and aims to keep products, components, and materials at their highest utility and value at all times".

A circular economy requires paradigmatic shifts in how we design, construct, and operate buildings, and in the way resources are managed within the built environment. This lecture will highlight both ongoing research projects of the Circular Construction Lab at Cornell University and built proto-typologies of 2hs Architects and Engineers from Germany in an attempt to trigger an informed discussion on the necessary and systemic changes to today's industry toward circular construction within a circular economy.

讲座介绍　Lecture introduction

随着气候变化、快速城市化和环境恶化，规划学科需要重新定位，以更好地服务和应对这些挑战。本讲座将讨论投射研究、实验教学法以及当前通向建筑和城市设计的相关设计方法——这些方法在"人类世"的时代以新的、相互关联的行动形式，推进各类策略性实践的整合。

演讲人：夏洛特·马尔泰尔·巴特　Speaker: Charlotte Malterre-Barthes

Faced with climate change, fast-paced urbanization, and environmental degradation everywhere, planning disciplines need to be repositioned to serve and address better these challenges. This lecture will discuss examples of projective research, experimental pedagogy, and currently relevant design approaches to architecture and urban design that promote the integration of strategic practices within new and interconnected forms of action in the time of the "Anthropocene".

第一次圆桌论坛　First roundtable forum
建构共生与未来环境建造
Constructing coexistence and futures of environment-making

第二次圆桌论坛　Second roundtable forum
建构共生与未来环境建造
Constructing coexistence and futures of environment-makining

研究生国际教学工作坊 POSTGRADUATE INTERNATIONAL DESIGN STUDIO

设计研究学导论
INTRODUCTION TO DESIGN RESEARCH

默里·弗雷泽　鲁安东
Murray Fraser, LU Andong

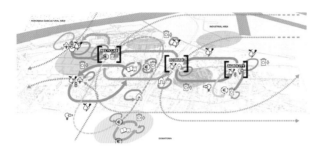

课程概述　Course overview

受2021年度校级重点引智项目"引进人文社科类资深海外专家重点支持计划"支持，伦敦大学学院默里·弗雷泽教授开设线上研究生课程"设计研究学导论"，课程聚焦当代设计研究的基本问题和前沿思想，将理论研究与设计实践对比印证，本专业选课学生共26人。

课程包括学术讲座系列、线上工作坊、文献精读3个版块。

1）学术讲座系列共3讲，仅限选课学生参加。

2）线上工作坊共进行了4次线上研讨，参加教学的嘉宾包括多位英国皇家建筑师学会安妮·斯宾克建筑教育终身成就奖得主、诸多国内外知名建筑师和理论家。线上直播听课总人数超过30 000人次。

3）配合课程开展文献精读计划，在全球范围进行招募并选拔出8名青年学人，与本校学生合作完成翻译任务。围绕翻译成果进行了2次汇报研讨会。

Supported by the "Key Support Program for the Introduction of Senior Overseas Experts in Humanities and Social Sciences", a key university-level intellectual attraction project in 2021, Professor Murray Fraser of University College London offered the online graduate course "Introduction to Design Research", which focuses on the fundamental issues and cutting-edge ideas of contemporary design research, and compares theoretical research with design practice, which was attended by 26 students.

The course included three sections: academic lecture series, online workshops, and intensive reading of literature:

1) The academic lecture series consisted of three lectures, which were restricted to students taking the course.

2) The online workshop series consisted of four online seminars with the participation of several RIBA Annie Spink Lifetime Achievement Award winners in architectural education and many famous architects and theorists from home and abroad. The total number of attendees of the live online workshops exceeded 30 000.

3) In conjunction with the course, a literature intensive reading program was conducted, and eight young scholars were recruited and selected from around the world to work with our students to complete the translation tasks. Two reporting seminars were held on the results of the translations.

学术讲座系列　Academic lecture series

设计研究天然带有艺术学、社会学、技术科学的跨学科的基因；设计研究要面向社会和政治背景，形成思辨与批判的方法；设计研究不仅是对已建成建筑的分析，更可以概念先行，用前瞻性的视角，通过对未来城市愿景的构想，反向塑造当代都市面貌。这三次讲座的目的是在全球范围内讨论有关建筑设计研究的问题。从总体概述开始，讲座讨论了与已建项目和推测性建议相关的设计研究，最后探讨该方法在未来如何变得更加全球化。这些讲座借鉴了巴特莱特设计研究作品集中的例子，展示如何构思和传播设计研究，并参考UCL出版社《建筑中的设计研究》系列丛书和《建筑研究领域杂志》（AJAR）作为传播设计研究的两种不同方式。

Design research has a natural interdisciplinary DNA of art, sociology, and technical science; design research should be oriented to social and political contexts and develop a discursive and critical approach; design research is not only the analysis of completed buildings, but can also be conceptual in nature, shaping the contemporary urban landscape in reverse through a forward-looking vision of future cities. The purpose of these three lectures is to discuss issues related to architectural research in a global context. Beginning with a general overview, the lectures discuss design research as it relates to built projects and speculative proposals, and conclude with an exploration of how the approach might become more global in the future. The lectures draw on examples from the Bartlett Design Research portfolio to demonstrate how design research can be conceived and disseminated, and refer to the UCL Press *Design Research in Architecture* series and the *Journal of Architectural Research Areas* (JARA) as two different ways of disseminating design research.

学术讲座1　Academic lecture 1

时间：2021年6月16日　人数：41人
Time: June 16, 2021　Number of people: 41

主题　Topic：
- 我们如何概念化建筑中不同类型的研究？
- 设计研究在哪些方面与建筑学的其他研究学科相关？
- 设计研究何时以及如何成为一门方法论？
- 我们现在从事设计研究的动机或压力是什么？

—How do we conceptualize different types of research in architecture?
—In what ways is design research related to other research disciplines in architecture?
—When and how did design research become a methodological discipline?
—What are our motivations or pressures to engage in design research now?

学术讲座2　Academic lecture 2

时间：2021年6月23日　人数：41人
Time: June 23, 2021　Number of people: 41

主题　Topic：
- 我们如何确定已建项目设计研究的议程/问题？
- 沟通这种特殊设计研究的最佳方式是什么？
- 我们如何通过设计研究分析已建项目之间的关系？
- 我们是否应该将"现实世界"条件视为有助于或阻碍设计研究？

—How do we define the agenda/questions of design research for established projects?
—What is the best way to communicate this particular design study?
—How do we analyze the relationships between built projects through design research?
—Should we consider "real world" conditions as helping or hindering design research?

学术讲座 3　Academic lecture 3
时间：2021 年 6 月 30 日　人数：41 人
Time: June 30, 2021　Number of people: 41
主题　Topic：
- 我们如何确定投机项目设计研究的议程 / 问题？
- 沟通这种特殊设计研究的最佳方式是什么？
- 我们是否应该将投机项目视为不同于设计研究的已建项目？
- 推测性设计研究能带来什么好处？
–How do we define the agenda/questions of design research for speculative projects?
–What is the best way to communicate this particular design study?
–Should we consider speculative projects as different from design studies as built projects?
–What benefits can speculative design studies provide?

线上工作坊系列（线上直播）　Online workshop series (live online)
　　工作坊共进行了四次线上直播，针对建筑学的未来，探讨"未来媒介"、交叉学科等议题。每次邀请一位设计研究著名理论家与两家进行设计研究实践的事务所，通过充分的理论陈述和实践展示，展开并深入探讨一个当代的设计研究议题。多家建筑媒体转载相关推送，在线观看总人数突破 3 万人。
The workshop consisted of four live webcasts, addressing the future of architecture, "future media", interdisciplinary issues etc. were discussed. Each time, a renowned design research theorist and two agencies conducting design research practices were invited to discuss a contemporary design research topic through a full theoretical presentation and practical demonstration. Many architectural media have republished the tweets, and the total number of online viewers exceeded 30,000.

圆桌 / 讲座系列 1（线上直播）　Roundtable/Lecture series 1 (live online)
时间：2021 年 7 月 7 日　最高在线人数：2 563 人
Time: July 7, 2021　Maximum number of people online: 2 563
主题：建筑设计研究的全球化　Topic: the globalization of architectural research
主讲人：默里·弗雷泽　Speaker: Murray Fraser
邀请嘉宾：莱斯利·洛科　Guest: Leslie Lokko

圆桌 / 讲座系列 2（线上直播）　Roundtable/Lecture series 2 (live online)
时间：2021 年 7 月 19 日　最高在线人数：22 565 人
Time: July 19, 2021　Maximum number of people online: 22 565
主题："未来 + 建筑 + 人类学"研讨会
Topic: "Future + Architecture + Anthropology" Seminar
主讲人：项飙、鲁安东、庄慎、童明
Speakers: XIANG Biao, LU Andong, ZHUANG Shen, TONG Ming
回应嘉宾：赵辰、李兴钢　Respondents: ZHAO Chen, LI Xinggang
主持人：默里·弗雷泽　Moderator: Murray Fraser

圆桌 / 讲座系列 3（线上直播）　Roundtable/Lecture series 3 (live online)
时间：2021 年 7 月 28 日　最高在线人数：2 687 人
Time: July 28, 2021　Maximum number of people online: 2 687
主题：东亚设计研究学术联盟"理论 + 实践对话"圆桌一
Topic: East Asia Design Research Consortium "Theory + Practice Dialogue" Roundtable 1
主讲人：梅兰莱·多德、源计划工作室　Speaker: Melanle Dodd, Source Project Studio
对话嘉宾：默里·弗雷泽　Guest: Murray Fraser

圆桌 / 讲座系列 4（线上直播）　Roundtable/Lecture series 4 (live online)
时间：2021 年 8 月 4 日　最高在线人数：2 890 人
Time: August 4, 2021　Maximum number of people online: 2 890
主题："理论 + 实践对话"圆桌二
Topic: "Theory + Practice Dialogue" Roundtable 2
城市更新中的设计研究　Design research in/for urban renewal
主讲人：乔纳森·希尔、梓耘斋建筑事务所、MUIR 建筑事务所、城村架构
Speaker: Jonathan Hill, Ziyunzhai Architects, MUIR Architects, Urban Village Architecture
对话嘉宾：默里·弗雷泽、王辉　Guest: Murray Fraser, WANG Hui
主持人：冯路　Moderator: FENG Lu

精读计划　Intensive reading
　　为加深学生对设计研究的理解，将理论联系实际，课程结合精读计划，提供 8 篇理论文章，由学生分组完成相关文章的翻译。随后安排了两次翻译成果的汇报研讨会，并邀请专家针对翻译过程中的概念或理论问题进行讨论、点评。预期出版《设计研究学理论读本》图书一部。
In order to deepen students' understanding of design research, the course combined with the intensive reading program to provide eight theoretical articles, which were translated by students in groups. The course is followed by two workshops to present the translation results. Experts are invited to discuss and comment on conceptual or theoretical issues in the translation process. The results of the translation are expected to be published as a book entitled *Readings in Design Research Theory*.

建筑设计课程
ARCHITECTURAL DESIGN COURSES

本科一年级
设计基础
鲁安东　唐莲　梁宇舒　尹航
课程类型：必修
学时学分：64学时／2学分

Undergraduate Program 1st Year
DESIGN FOUNDATION · LU Andong, TANG Lian, LIANG Yushu, YIN Hang
Type: Required Course
Study Period and Credits: 64 hours/2 credits

课程内容
　　本教案基于四条主题线索和三个能力培养阶段，设计了12个时长五周的教学模块。学生可以根据自己的兴趣和需求自由选修不同模块，量身塑造自己的设计思维和设计能力。通过模块化教学，本教案发挥了通识教育下自主学习的优势，开展理性、全面的思维训练，突出系统、多元的能力培养。通过将设计基础作为创意工科的"元"学科，既为学生进一步的专业学习打下扎实基础，也培养了学生未来跨学科创新的必要素质。

Course content
In this teaching plan, 12 five-week teaching modules are designed based on the four thematic clues and three competence training stages. Students can freely select different modules according to their own interest and needs, to tailor their design thinking and design competence. Through modular teaching, this teaching plan can make use of the advantages of autonomous learning under liberal education, thus carrying out rational and comprehensive thinking training, and highlighting systematic and multivariate competence training. Design foundation, the "meta" subject of creative engineering, can lay a solid foundation for professional learning by the students, and cultivate the necessary competences for interdisciplinary innovation in the future.

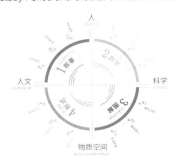

本科二年级
建筑设计基础
刘铨　史文娟
课程类型：必修
学时学分：64学时／4学分

Undergraduate Program 2nd Year
ARCHITECTURAL DESIGN FOUDATION · LIU Quan, SHI Wenjuan
Type: Required Course
Study Period and Credits: 64 hours/4 credits

课程内容
　　在重新认识建筑基础知识的前提下，将认知与表达作为这门课的教学主线，依照循序渐进的原则，分三个阶段设置了不同的教学任务，每个阶段有其特定的认知对象和认知方法，，包含着若干练习同时每个阶段的训练都建立在之前一个阶段学习要点的基础上，力图更好地使学生通过认知的过程从一个外行逐步进入专业领域，并为后续的建筑设计学习打下宽阔和扎实的基础。

阶段一：
建筑立面局部测绘：从材料与构件尺寸认知到正投影图绘制；
建筑物测绘：从建筑空间分割与功能尺度认知到平、剖面图绘制；
建筑窗测绘：从建筑构造认知到大样图绘制。
阶段二：
建筑结构模型制作：从结构图识图到结构模型制作；
墙身模型制作：从大样图识图到构造模型制作。
阶段三：
街道空间认知：理解街巷肌理、城市街道空间及其限定与功能；
地块与建筑类型认知：理解地块肌理、城市建筑类型及其功能与交通组织；
地形与气候认知：理解自然地形、植被及日照等自然环境要素。

Course content
Under the premise of re-understanding the basic knowledge of architecture, cognition and expression are taken as the main teaching line of this course. According to the principle of step-by-step, different teaching tasks are set in three stages. Each stage has its specific cognitive objects and cognitive methods, including several exercises. At the same time, the training of each stage is based on the learning points of the previous stage, trying to better enable students to gradually enter the professional field from a layman through the cognitive process, and lay a broad and solid foundation for subsequent architectural design learning.

Stage 1:
Local surveying and mapping of building facade: From recognizing the size of materials and components to drawing orthographic projection.
Building surveying and mapping: Recognizing the plan and section drawing from building space segmentation and functional scale.
Building window surveying and mapping: From building structure cognition to detail drawing.
Stage 2:
Building structural model making: From structural drawing recognition to structural model making.
Wall body model making: from detail drawing identification to structural model making.
Stage 3:
Street space cognition: Understanding street texture, urban street space and its limitations and functions.
Cognition of the plot and building type: Understanding plot texture, urban building type, and its function and traffic organization.
Terrain and climate cognition: Understanding natural environment elements such as natural terrain, vegetation and sunshine.

本科二年级

建筑设计（一）：独立居住空间设计
· 刘铨　冷天　吴佳维
课程类型：必修
学时学分：64 学时 / 4 学分

Undergraduate Program 2nd Year
ARCHITECTURAL DESIGN 1: INDEPENDENT LIVING SPACE DESIGN · LIU Quan, LENG Tian, WU Jiawei
Type: Required Course
Study Period and Credits: 64 hours/4 credits

课程内容

本次练习的主要任务是综合运用前期案例学习中的知识点——建筑在水平方向上如何利用高度、开洞等操作划分空间，内部空间的功能流线组织及视线关系，墙身、节点、包裹体系、框架结构的构造方式，周围环境对空间、功能、包裹体系的影响等，初步体验一个小型独立居住空间的设计过程。

教学要点

1. 场地与界面：本次设计的场地面积在 80—100 m²，场地单面或相邻两面临街，周边为 1—2 层的传统民居。
2. 功能与空间：本次设计的建筑功能为小型家庭独立式住宅（附设有书房功能）。家庭主要成员包括一对年轻夫妇和 1—2 位儿童（7 岁左右）。新建建筑面积 160—200 m²，建筑高度 ≤ 9 m（不设地下空间）。设计者根据设定的家庭成员的职业及兴趣爱好确定空间的功能（职业可以是但不局限于理、工、医、法的技术人员）。
3. 流线组织与出入口设置：考虑建筑内部流线合理性以及建筑出入口与场地周边环境条件的合理衔接。
4. 尺度与感知：建筑中的各功能空间的尺寸需要以人体尺度及人的行为方式作为基本的参照，并通过图示表达空间构成要素与人的空间体验之间的关系。

Course content

The main task of this exercise is to comprehensively use the knowledge points in the early case study—how to use height, opening and other operations to divide space in the horizontal direction of the building, functional streamline organization and line of sight relationship of internal space, construction mode of wall body, nodes, wrapping system and frame structure, the influence of the surrounding environment on the space, function and wrapping system, and preliminarily experience the design process of a small independent living space.

Teaching Essential

1. Site and interface: The site of this design covers an area of about 80–100 m², facing the street on one side or two adjacent sides, surrounded by 1–2 floors of traditional residential buildings.
2. Function and space: The building function of this design is a small family independent residence (with study function attached). The main members of the family include a young couple and 1–2 children (about 7 years old). The new building area is 160–200 m² and the building height is ≤ 9 m (no underground space). The designer determines the function of the space according to the set occupation and interests of family members (the occupation can be but not limited to technicians of science, engineering, medicine and law).
3. Streamline organization and entrance and exit setting: Consider the rationality of the internal streamline of the building and the reasonable connection between the entrance and exit of the building and the surrounding environmental conditions of the site.
4. Scale and perception: The size of each functional space in the building needs to take the human body scale and human behaviors as the basic reference, and express the relationship between spatial constituent elements and human spatial experience through diagrams.

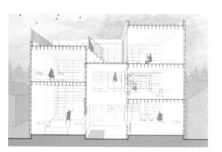

本科二年级

建筑设计（二）：文怀恩旧居加建设计
· 刘铨　冷天　吴佳维
课程类型：必修
学时学分：64 学时 / 4 学分

Undergraduate Program 2nd Year
ARCHITECTURAL DESIGN 2: DESIGN FOR EXTENSION OF WEN HUAI'EN'S FORMER RESIDENCE · LIU Quan, LENG Tian, WU Jiawei
Type: Required Course
Study Period and Credits: 64 hours/4 credits

课程内容

1. 功能要求：根据文怀恩与金陵大学的历史，设计一个展示纪念馆。老建筑由于缺少大空间，需要加设一个较大的灵活空间。建成后老建筑用作固定展陈和办公，新建筑则作为临时展览、研讨交流、会议茶歇等可以灵活使用的空间。同时由于其位于校园核心地带，新增建筑内拟设一个小型咖啡厅，为学校教职工及日常参观人群提供服务。新建建筑总面积不少于 150 m²，建筑高度 ≤ 8 m（檐口高度，不包括女儿墙）。新建建筑以一层为主，局部可设夹层。在场地内还应考虑一处与展览主题相关的纪念性空间。新老建筑应作为一个整体考虑其参观流线，但新建建筑也应考虑其相对独立性，在老建筑闭馆时也可独立使用。
2. 场地环境：现状建筑和场地内各项要素既是限制，又是形成新建筑体量的基本条件。本次设计场地南侧为小礼拜堂，北侧为教学楼，东侧面对小花园（原金陵大学主轴线上的入口花园）和金陵大学主轴线上的石碑、雕塑。结合与这些重要的环境要素的位置的视觉关系来考虑建筑物、纪念性空间的布局以及参观流线的组织。
3. 空间限定要素与视觉关系的组织：本次训练需要通过空间限定要素（水平与垂直构件）与身体感知（路径、视线、活动与尺度、光影、质感）关系的组织，塑造出相应的室内外展陈与纪念性空间，连接文宅历史记忆与现实需求，创造性地再现该场所的人文内涵。
4. 材料与建造：选择合适的材料、结构形式，呼应空间组织与场地环境需要。

Course content

1. Functional requirements: Design an exhibition memorial hall according to the history of Wen Huai'en and Jinling University. Due to the lack of large space in the old building, a large flexible space needs to be added. After completion, the old building is used for fixed exhibition and offices, while the new building is used as a flexible space for the temporary exhibition, discussion and exchange, conference and tea break.
2. Site environment: The current buildings and various elements in the site are not only restrictions, but also the basic conditions for the formation of new building volume. The south side of the design site is the small chapel, the north side is the teaching building, and the east side faces the small garden (the entrance garden on the original Jinling University main axis) and the stone tablets and sculptures on the Jinling University main axis. Combined with the visual relationship with the location of these important environmental elements, the layout of buildings, memorial spaces and the organization of visiting streamlines are considered.
3. Organization of space limiting elements and visual relationship: This training needs to shape the corresponding indoor and outdoor exhibition and memorial space through the organization of the relationship between space limiting elements (horizontal and vertical components) and body perception (path, line of sight, activity and scale, light and shadow, texture), connect the historical memory and practical needs of the Wen house, and creatively reproduce the humanistic connotation of the place.
4. Materials and construction: Select appropriate materials and structural forms to meet the needs of space organization and site environment.

本科三年级
建筑设计（三）：专家公寓设计
华晓宁　窦平平　黄华青
课程类型：必修
学时学分：72学时／4学分

Undergraduate Program 3rd Year
ARCHITECTURAL DESIGN 3: THE EXPERT APARTMENT DESIGN · HUA Xiaoning, DOU Pingping, HUANG Huaqing
Type: Required Course
Study Period and Credits: 72 hours/4 credits

课程内容
　　拟在南京大学鼓楼校区南园宿舍区内新建专家公寓一座，用于国内外专家到访南大开展学术交流活动期间的居住。用地位于南园中心喷泉西侧，面积约3 600 m²。地块上原有建筑将被拆除，新建筑总建筑面积不超过3 000 m²。高度不超过3层。

教学目标
　　从空间单元到系统的设计训练。
　　从个体到整体，从单元到体系，是建筑空间组织的一种基本和常用方式。本课题首先关注空间单元的生成，并进一步根据内在的使用逻辑和外在的场地条件，将多个单元通过特定方式与秩序组合起来，形成一个兼备合理性、清晰性和丰富性的整体系统。基本单元的重复、韵律、变异等都是常用的操作手法。

Course content
It is proposed to build a new expert apartment in the Nanyuan dormitory area of Gulou campus of Nanjing University for domestic and foreign experts to live during their visit to Nanjing University for academic exchange activities. The land is located in the west of the fountain in the center of Nanyuan, covering an area of about 3 600 m². The original buildings on the plot will be demolished, and the total construction area of the new buildings will not exceed 3 000 m². The height shall not exceed 3 floors.

Teaching objectives
From space unit to system design training.
From individual to whole, from unit to system, is a basic and common way of architectural space organization. This topic first pays attention to the generation of spatial units, and further combines multiple units with order in a specific way according to the internal use logic and external site conditions to form an overall system with rationality, clarity and richness. Repetition, rhythm and variation of basic units are commonly used.

本科三年级
建筑设计（四）：世界文学客厅
华晓宁　窦平平　黄华青
课程类型：必修
学时学分：72学时／4学分

Undergraduate Program 3rd Year
WORLD ARCHITECTURAL DESIGN 4: LITERATURE LIVING ROOM · HUA Xiaoning, DOU Pingping, HUANG Huaqing
Type: Required Course
Study Period and Credits: 72 hours/4 credits

课程内容
　　南京古称金陵、白下、建康、建邺……历来是人文荟萃、名家辈出之地，号称"天下文枢"。南京作为六朝古都，亦为中国文学之始。何为文？梁元帝曰："吟咏风谣，流连哀思者，谓之文。"汉魏有文无学，六朝文学《文选》《文心雕龙》《诗品》既是文学评论的开始，也是文学的发端。
　　2019年，南京入选联合国"世界文学之都"，开展一系列城市空间计划，包括筹建"世界文学客厅"，作为一座以文学为主题的综合性博物馆。该馆选址位于北极阁公园东南隅，用地面积约5 050 m²，紧临市政府中轴线，毗邻古鸡鸣寺、玄武湖、明城墙、东南大学四牌楼校区等历史文化遗迹，构成城市与山林之间的过渡空间。设计应妥善处理建筑与周边城市环境和既有建筑的关系，彰显中国文学的精神特质。

教学目标
　　本课程主题是"空间"，学习建筑空间组织的技巧和方法，训练对空间的操作与表达。空间问题是建筑学的基本问题。本课题基于文学主题，训练文本、叙事与空间序列的串联，学习空间叙事与空间用途的整体构思，充分考虑人在空间中的行为、空间感受，尝试以空间为手段表达特定的意义和氛围，最终形成一个完整的设计。

Course content
In 2019, Nanjing was selected as the "capital of world literature" of the United Nations and plans to carry out a series of urban space plans, including preparing to build the "world literature living room" as a comprehensive museum with literature as the theme. The museum is located in the southeast corner of Beijige Park, with a land area of about 5 050 m². It is close to the central axis of the municipal government and adjacent to historical and cultural sites such as ancient Jiming Temple, Xuanwu Lake, Ming City Wall and Sipailou campus of Southeast University, forming a transition space between the city and mountains. The design should properly deal with the relationship between the building and the surrounding urban environment and existing buildings, and highlight the spiritual characteristics of Chinese literature.

Teaching objectives
The theme of this course is "space", learning the skills and methods of architectural space organization, and training the operation and expression of space. Space problem is the basic problem of architecture. Based on the literary theme, this topic trains the series of the text, narration and spatial sequence, learns the overall idea of spatial narration and spatial use, fully considers people's behaviors and spatial feelings in space, tries to express specific meaning and atmosphere by means of space, and finally forms a complete design.

本科三年级
建筑设计（五）：大学生健身中心改扩建设计·傅筱 钟华颖 王铠
课程类型：必修
学时学分：64学时 / 4学分

Undergraduate Program 3rd Year
ARCHITECTURAL DESIGN 5: RECONSTRUCTION AND EXPANSION DESIGN OF COLLEGE STUDENT FITNESS CENTER · FU Xiao, ZHONG Huaying, WANG Kai
Type: Required Course
Study Period and Credits: 64 hours/4 credits

课程内容

将结构训练的重点放在结构、形态与空间的关联性上。在以前的课程设计中学生较为熟悉的是框架、砖混结构，虽然每一次课程设计均与结构不可分割，但是结构被外围保护材料包裹，在建筑表达上始终处于被动"配合"的状态。对于有一定跨度的空间，容易让学生建立一种"主动"运用结构的意识。为了达到这个训练目标，教学团队认为跨度不宜太大，以 30 m 左右为宜，结构选型余地较大，结构与空间的配合受到技术条件的制约相对较小，以利于学生充分理解空间语言与结构的关联。

强调建筑设计训练的综合性，结构只是一个重要的要素，即不过分放大结构的作用，在将结构作为空间生成的一种推动力的同时要求学生综合考虑场地、使用、空间感受、采光和通风等基本要素。实际上通过短短 8 周的教学就希望学生能够达到实际操作层面的综合性是不现实的。教学的目标是让学生建立起综合性的认知，所以题目设置宜简化使用功能，明确物理性能要求，给定设备和辅助空间要求，将琐碎的知识点化为知识模块给定学生，让学生更多地学会模块间的组织，而不需要确切了解其中的每一个技术细节。

结构选型是指定的，但鼓励学生结合性能需求进行合理改变；采光、通风要求是明确的，但鼓励学生结合人的需求进行设计；设备和辅助空间是给定的，但要求学生学会在空间上进行合理布置，理解只有设备和辅助空间布置妥当才能创造使用空间的价值。

Course content

The focus of structural training is on the relationship between structure, form and space. In the previous curriculum design, students were familiar with the frame and brick concrete structure. Although each curriculum design was inseparable from the structure, the structure was wrapped by external enclosure materials and was always in a passive "cooperation" state in architectural expression. For a space with a certain span, it is easy for students to establish a sense of "active" use of the structure. In order to achieve this training goal, the teaching team believes that the span should not be too large. It is appropriate to be about 30 m. There is a large room for structural selection, and the coordination between structure and space is relatively less restricted by technical conditions, so as to help students fully understand the relationship between spatial language and the structure.

It emphasizes the comprehensiveness of architectural design training, and the structure is only an important element. That is to say, the role of structure is not enlarged. While taking the structure as a driving force for space generation, students are required to comprehensively consider the basic elements such as the site, use, space feeling, daylighting and ventilation. In fact, it is unrealistic to hope that students can achieve the comprehensiveness of practical operation through just 8 weeks of teaching. The goal of teaching is to enable students to establish a comprehensive cognition. Therefore, the topic setting should simplify the use function, clarify the requirements of physical performance, offer equipment and auxiliary space requirements, and turn trivial knowledge points into knowledge modules. Students learn more about the organization between modules without knowing every technical detail.

本科三年级
建筑设计（六）：社区文化艺术中心设计
张雷 钟华颖 王铠
课程类型：必修
学时学分：64学时 / 4学分

Undergraduate Program 3th Year
ARCHITECTURAL DESIGN 6: DESIGN OF COMMUNITY CULTURE AND ART CENTER · ZHANG Lei, ZHONG Huaying, WANG Kai
Type: Required Course
Study Period and Credits: 64 hours/4 credits

课程内容

拟在百子亭风貌区基地处新建社区文化中心，总建筑面积约为 8 000 m^2，项目不仅为周边居民提供文化基础设施，同时也期望成为复兴老城的街区活力的文化地标。根据基地条件、功能使用进行建筑和场地设计。总用地详见附图，基地用地面积为 4 600 m^2。

设计内容

1. 演艺中心：包含 400 座小剧场，乙级。台口尺寸为 12 m×7 m。根据设计的等级确定前厅、休息厅、观众厅、舞台等面积。观众厅主要为小型话剧及戏剧表演而设置。按 60—80 人化妆布置化妆室及服装道具室，并设 2—4 间小化妆室。要求有合理的舞台及后台布置，应设有排练厅、休息室、候场区以及道具存放间等设施，其余根据需要自定。

2. 文化中心：定位于区级综合性文化站，包括公共图书阅览室、电子阅览室、多功能厅、排练厅以及辅导培训、书画创作等功能室（不少于 8 个且每个功能室面积应不低于 30 m^2）。

3. 配套商业：包含社区商业以及小型文创主题商业单元。其中社区商业为不小于 200 m^2 超市一处，文创主题商业单元面积为 60—200 m^2。

4. 其他：变电间、配电间、空调机房、售票、办公、厕所等服务设施根据相关设计规范确定，各个功能区可单独设置，也可统一考虑。地上不考虑机动车停车配建，街区地下统一解决，但需要根据建筑功能面积计算数量。

Course content

The project plans to build a new community culture and art center at the base of Baiziting historic area, with a total construction area of about 8 000 m^2. The project not only serves the cultural infrastructure of surrounding residents, but also hopes to become a cultural landmark to revive the vitality of the old city. Design the building and site according to the base conditions and functional use. The total land is shown in the attached figure, and the land area of the base is 4 600 m^2.

Design Content

1. Performing arts center: It contains 400 small theatres, class B. The size of the proscenium is 12 m × 7 m. Determine the area of front hall, lounge, auditorium and stage according to the design level.
2. Cultural center: Located at the district level comprehensive cultural station, it includes public book reading room, electronic reading room, multi-functional hall, rehearsal hall, counseling and training, calligraphy and painting creation and other functional rooms.
3. Supporting business: It includes community business and small cultural and creative theme business units. Among them, the community business is a supermarket with an area of no less than 200 m^2, and the area of cultural and creative theme business unit is 60-200 m^2.
4. Others: Service facilities such as substation room, power distribution room, air conditioning room, ticketing, office and toilet are determined according to relevant design specifications.

本科四年级
建筑设计（七）：高层办公楼设计
吉国华　胡友培　尹航
课程类型：必修
学时学分：64 学时 / 4 学分

Undergraduate Program 4th Year
ARCHITECTURAL DESIGN 7: DESIGN OF HIGH-RISE OFFICE BUILDINGS
· JI Guohua, HU Youpei, YIN Hang
Type: Required Course
Study Period and Credits: 64 hours/4 credits

教学目标

生态性能驱动的办公建筑设计涉及城市、空间、形体、环境、能耗、结构、设备、材料、消防等方面内容，是一项较复杂与综合的任务。有效的空间组织、适应性形体、交互性表皮以及性能化结构设计等策略，对建筑室内外环境的生态性能起着决定性的作用。本课题教学重点和目标是帮助学生理解、消化以上知识，提高综合运用并创造性解决问题的技能，学习并运用生态性能模拟分析软件，以生态性能驱动建筑设计。

设计内容

1. 经济技术指标与场地
用地面积为 4 520 m²，地上总建筑面积 ≥ 35 000 m²，建筑限高 ≤ 100 m。
2. 功能要求
办公：设计应兼顾各种办公空间形式。
会议：须设置 400 人报告厅 1 个、200 人报告厅 2 个、100 人报告厅 4 个，其他各种会议形式的中小型会议室若干，以及咖啡 / 茶室、休息厅、服务用房等。
机动车交通：机动车交通独立设置，人车分离。场地交通流线，须结合现状周边情况，统一考虑。地下部分为车库和设备用房。

Teaching objectives
The design of office buildings driven by eco-performance involves the aspects of the city, space, form, environment, energy consumption, structure, equipment, materials, and fire protection. It is a complex and comprehensive task. The strategies such as effective spatial organization, adaptive shapes, interactive surfaces, and performance-based structural design play a decisive role in ecological performance of indoor and outdoor environment. This course intends to help the students to understand and digest the knowledge of various aspects, improve comprehensive application and creative problem solving skills, learn and use the ecological performance simulation analysis software, and drive architectural design with ecological performance.

Design content
1. Economic and technical indicators and site
Land area: 4 520 m², the total building area above ground 35 000 m², height limit ≤ 100 m.
2. Functional requirements
Office: The design should take into account various forms of office space.
Meeting: There should be a 400-person conference hall, two 200-person conference halls, and four 100-person conference halls. There should also be several small and medium-sized conference rooms in various forms, and the cafe/tea bar, lounge, and service room.
Vehicle traffic: The vehicle traffic should be set independently, with separation between people and vehicles. Site traffic flow should be considered in combination with the surrounding situations. The underground part consists of the garage and equipment room.

本科四年级
建筑设计（八）：城市设计
童滋雨　尹航　尤伟
课程类型：必修
学时学分：64 学时 / 4 学分

Undergraduate Program 4th Year
ARCHITECTURAL DESIGN 8: URBAN DESIGN
· TONG Ziyu, YIN Hang, YOU Wei
Type: Required Course
Study Period and Credits: 64 hours/4 credits

课程内容
计算化城市设计

教学目标

中国的城市发展已经逐渐从增量扩张转向存量更新。通过对城市建成环境的更新改造而提升环境性能和质量，将成为城市建设的新热点和新常态。与此同时，5G、物联网、无人驾驶等技术的发展又给城市环境的使用方式带来了新的变化。如何在城市更新设计中拓展建筑设计的边界也就成为新的挑战。

城市更新不但需要对建成环境本身有更充分的认知，也要对其中的人流、车流乃至水流、气流等各种动态的活动有正确的认知。从设计上来说，这也大大提高了设计者所面临的问题的复杂性，仅靠个人的直观感受和形式操作难以保证设计的合理性。而借助空间分析、数据统计、算法设计等数字技术，我们可以更好地认知城市形态的特征，理解城市运行的规则，并预测城市未来的发展。通过规则和算法来计算生成城市也是对城市设计思维范式的重要突破。

因此，本次设计将针对这些发展趋势，以城市街巷空间为研究对象，通过思考和推演探索其更新改造的可能性。通过本次设计，学生们可以理解城市设计的相关理论和方法，掌握分析城市形态和创造更好城市环境质量的方法。

Course content
Computational urban design

Teaching objectives
China's urban development has gradually shifted from incremental expansion to stock renewal. Improving environmental performance and quality through the renewal and transformation of urban built environment will become a new hot spot and new normal of urban construction. At the same time, the development of 5G, Internet of things, unmanned driving and other technologies has brought new changes to the use of urban environment. How to expand the boundary of architectural design in urban renewal design has become a new challenge.
Urban renewal not only needs to have a better understanding of the built environment itself, but also has a correct understanding of the people stream, vehicles stream, water stream, air stream and other dynamic activities. In terms of design, it also greatly improves the complexity of the problems faced by designers. It is difficult to ensure the rationality of design only by personal intuitive feeling and formal operations. With the help of various digital technologies such as spatial analysis, data statistics and algorithm design, we can better understand the characteristics of the urban form, understand the rules of the urban operation, and predict the future development of the city. Calculating and generating cities through rules and algorithms is also an important breakthrough in the thinking paradigm of urban design.
Therefore, this design will aim at these development trends, take the urban street space as the research object, and explore the possibility of its renewal and transformation through thinking and deduction. Through this design, students can understand the relevant theories and methods of urban design, and master the methods of analyzing the urban form and creating better urban environmental quality.

本科四年级
本科毕业设计
童滋雨 李清朋 吉国华
课程类型：必修
学时学分：1 学期 /0.75 学分

Undergraduate Program 4th Year
GRUDUATION PROJECT · TONG Ziyu, LI Qingpeng, JI Guohua
Type: Required Course
Study Period and Credits: 1 term /0.75 credit

课程内容
基于规则和算法的设计和搭建

本课题初步设定是设计一个具有一定跨度和高度的构筑物，该构筑物可通过砌块或编织等方法构建，能够容纳 3 人左右的活动空间。要求空间适宜，结构合理，形体的生成应具有相应的几何规则和算法，并通过数字化加工完成最终的模型搭建。

基于力学生形的数字化设计与建造

本课题要求学生在学校自选环境中设计一处用地面积 4 m × 4 m、遮盖面积为 10 m² 左右的建筑空间，以满足师生停留、休憩、交流的功能需求。课程通过实物模型制作不断探索设计问题。用数字化的方法研究和解决问题，最终通过数控加工的方式实现具有真实细节的构筑物。

Course content
Design and construction based on rules and algorithms
The preliminary setting of this topic is to design a structure with a certain span and height, which can be constructed by means of block or weaving, and can accommodate an activity space for about 3 people. It is required that the space is suitable and the structure is reasonable. The generation of the shape should have corresponding geometric rules and algorithms, and the final model construction should be completed through digital processing.

Digital design and construction based on mechanical form
The topic requires students to design a building space with a land area of 4 m × 4 m, and covering an area of about 10 m² in the school's optional environment, so as to meet the functional needs of teachers and students to stay, rest and communication. The topic continues to explore the design problem through the production of the physical models. The problem is studied and solved by digital methods, and finally the structure with real details is realized by NC machining.

本科四年级
本科毕业设计
黄华青 冉光沛
课程类型：必修
学时学分：1 学期 /0.75 学分

Undergraduate Program 4th Year
GRUDUATION PROJECT · HUANG Huaqing, RAN Guangpei
Type: Required course
Study Period and Credits: 1 term /0.75 credit

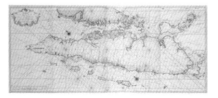

课程内容
流动的空间：海上贸易的技术传播与东南亚港口城市的近代塑造

教学内容
空间的"能动性"是建筑学、人类学等多学科关注的话题，体现于空间形式的自治性和空间—社会的互动建构，进而成为跨越主体与客体、个体与集体、想象与实体研究的桥梁。本研究从列斐伏尔、布迪厄、白馥兰的现代空间理论脉络出发，探讨空间作为一种兼具物质性/社会性的"技术"传播媒介，如何作为"能动者"构成并形塑近现代城市聚落的物质和社会景观。

本课题在"一带一路"、聚落文化、近代遗产等视野下，以 15—19 世纪兴盛的海上贸易沿线港口城市为载体，探讨技术（包括建筑、机械等物质技术及人力、资本、组织等社会技术）的传播与东南亚、南亚近现代港口城市的形成与变迁之间的激烈互动。

本课题基于史料挖掘和田野调查，从聚落形态和建筑类型出发，结合全球史、社会经济史、社区史等视角，为东南亚近代港口城市的空间形塑寻找以"人"为基准的线索，并探寻中国在这一流动的物质、社会、文化网络中发挥的作用。

Course content
Space for flow: the technological diffusion of maritime trade and the modern shaping of southeast asian port cities

Teaching content
The "agency" of space is a topic concerned by many disciplines such as architecture and anthropology, which is reflected in the autonomy of space form and the interactive construction between space and society, and then becomes a bridge across subject and object, individual and collective, imagination and entity. This study starts from H. Lefebvre, P. Bourdieu and F. Bray's context of modern space theory to discuss how space, as a "technical" media with both material and social characteristics, constitutes and shapes the material and social landscape of modern urban settlements as an "agent".
From the perspective of "one belt, one road", the settlement culture and the modern heritage, also based on the port city that flourished along the 15–19 century, this paper discusses the fierce interaction between the dissemination of technology (including material technology such as construction and machinery, and social technology such as human, capital and organization) and the formation and change of modern port cities in Southeast Asia and South Asia.
Based on historical data mining and field investigation, starting from settlement forms and architectural types, combined with the perspectives of global history, socio-economic history and community history, this topic seeks clues based on "people" for the spatial shaping of modern port cities in Southeast Asia, and explores the role of China in this mobile material, social and cultural network.

研究生一年级
建筑设计研究（一）：基本设计
傅筱
课程类型：必修
学时学分：40 学时 / 2 学分

Graduate Program 1st Year
ARCHITECTURAL DESIGN STUDIO 1: BASIC DESIGN
• FU Xiao
Type: Required Course
Study Period and Credits: 40 hours/2 credits

课程内容
建筑要求：
1）按照南京市宅基地相关规定，建筑基底面积不得超过 130 m²。要求内部布局紧凑经济，使用功能合理，在满足功能需求之下，尽量减少面积以省造价，总建筑面积不得超过 210 m²。
2）可选用结构为：钢筋混凝土框架结构、砖混结构、轻钢龙骨结构体系、木框架结构体系（同一地块的小组不得选用相同的结构体系）。外墙材料选择需与结构体系有一定的关联性，并考虑保温隔热要求。
3）入户空间要求朝南或者朝东。
4）空调形式为分体挂机或柜机，需设计放置位置。
5）明厨明卫。
6）具体房间数量要求：
（1）客厅；（2）餐厅；（3）厨房；（4）客卧 1 间（带卫生间）；（5）主卧 1 间（带卫生间）；（6）次卧 1 间（使用公用卫生间）；（7）画室 1 间（使用面积不小于 30 m²）；（8）公用卫生间，根据需要确定数量；（9）储藏空间。
技术要求：鼓励用 BIM（Revit）、Enscape 设计和表达。
参考书目：《加拿大木框架房屋建筑》（学院资料室）。

Course content
Building Requirements:
1) In accordance with the relevant regulations of Nanjing City residential base, the building base area shall not exceed 130 m². The internal layout is compact and economical, the usage function is reasonable, and the area is reduced as much as possible to save the construction cost under the satisfaction of the function demand. The total building area shall not exceed 210 m².
2) The optional structures are: reinforced concrete frame structure, brick concrete structure, light steel keel structure system, wood frame structure system (the same plot of the group shall not choose the same structural system). The choice of exterior wall materials should be related to the structural system and should consider the heat insulation requirements.
3) The entrance space should face south or east.
4) The air conditioner is in the form of split hanging machine or cabinet machine, and the placement position should be designed.
5) Bright kitchen and bright bathroom.
6) The number of specific rooms requires that:
(1) Living room; (2) Dining room; (3) Kitchen; (4) 1 guest bedroom (with bathroom); (5) 1 master bedroom (with bathroom); (6) 1 room of secondary bedroom (using common bathroom); (7) 1 drawing room (use area not less than 30 m²); (8) Common bathroom, the number is determined according to the need; (9) Storage space.
Technical requirements: BIM (Revit), Enscape design and expression are encouraged.
Bibliography: Canadian Wood Frame House Architecture (College Resource Room).

研究生一年级
建筑设计研究（一）：基本设计
金鑫
课程类型：必修
学时学分：40 学时 / 2 学分

Graduate Program 1st Year
ARCHITECTURAL DESIGN STUDIO 1: BASIC DESIGN
• JIN Xin
Type: Required Course
Study Period and Credits: 40 hours/2 credits

课程内容
1. 研究问题：南京大学建筑学院大学文科楼原本并非专门为建筑学所设计的教学空间，随着时间的推移，建筑学院对使用空间又提出新的要求。面对更新，并非一味推倒重来。如何既解决空间使用问题，又能保留不同时期的历史痕迹，还能创造新的可能，是此次课程所研究的。
2. 功能重组：从空间功能上，不再单纯是从前单一的教室功能，教学场所更加多元开放。教师和学生的行为模式发生变化，互动交流需求的增加，需灵活多变、模式各异的交流空间。此外还有展示空间、实验室空间、模型制作空间等新功能的提出，都会对建筑的水平空间与垂直空间提出新的要求。
3. 形式"外卷"：改造空间中运用多样的形式语言，能较为直观地反映空间属性，包括结构形式、围护形式、装饰元素等。同时要求一定程度上保留建筑空间中不同时期的形式，给使用者感知改造的时间性。
4. 研究目的：遗产保护和建筑考古学中有"地层学"这个概念，但在建筑改造设计中，更多的是一种选择性的形式表达。这个课程，通过对建筑环境和单体的历史研究，选择性地对不同时期的改造形式进行保留，同时对建筑空间形式进行创新，以此引发学生对与建筑改造设计的深入思考。

Course content
1. Research problem: the School of Architecture puts forward new requirements for the use of the Liberal Arts Building. How to not only solve the problem of space use, but also preserve the historical traces of different periods and create new possibilities, is the study of this course.
2. Function reorganization: teaching places are more diversified and open. The demand for interactive communication has increased, and there is a need for flexible and changeable communication space with different modes.
3. The form "outer volume": the use of various formal languages in the transformation of the space can more intuitively reflect the attributes of the space, including structural forms. At the same time, it is required to retain the forms of different periods in the architectural space to a certain extent, so as to give users a perception of the timeliness of the transformation.
4. Research purpose: this course, through the historical research of the built environment and individual units, selectively preserves the forms of renovation in different periods, and at the same time innovates the form of architectural space, so as to arouse students' in-depth thinking about architectural renovation and design.

研究生一年级
建筑设计研究（一）：概念设计
鲁安东
课程类型：必修
学时学分：40 学时 / 2 学分

Graduate Program 1st Year
ARCHITECTURAL DESIGN STUDIO 1: CONCEPTUAL DESIGN · LU Andong
Type: Required Course
Study Period and Credits: 40 hours/2 credits

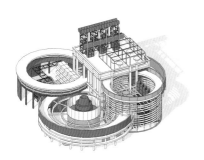

课程内容

1. 研究问题：增强场所

如何理解当代城市的日常空间？传统意义上的非物质要素正变得可感可触，无处不在的技术正日益成为人的基本能力的外延，并决定了人与世界的基本交互的形式，超物的在地显现使得每一个场所都无法仅作为它自身被认知……这些都构成了场所真实性的一部分，而场所的物质维度则成为或回归于它应有的"中介"本质。现代建筑对于独立存在的物质维度的预设以及在此基础上发展出的大量设计手段和设计价值是否依然有效？我们如何真正地为人设计？我们如何为真正的人设计？

2. 载体对象：文学之都

2019 年南京获得联合国教科文组织的"世界文学之都"称号。"世界文学之都"城市空间计划(2020—2023)是在"全球创意城市网络"的国际化语境下，多维度运用文学资源为城市赋能的行动计划。本课题将在此实际项目框架下，探索在地化设计如何一方面为本地赋能，另一方面充分运用增强场所的新行动机遇发挥更大的作用。

Course content

1. Research question: enhancing place

How to understand the daily space of the contemporary city? Traditionally immaterial elements are becoming palpable, ubiquitous technology is increasingly becoming an outgrowth of basic human capacities and determining the form of basic human interaction with the world, and the local manifestation of the supra-physical makes every place impossible to be perceived only as itself...These form part of the reality of place, and the material dimension of place becomes or returns to its proper "mediating" nature. Is the modern architectural presupposition of a separate material dimension, and the vast array of design tools and design values that have developed on this basis, still valid? How do we truly design for people? How do we design for real people?

2. The carrier object: the capital of literature

In 2019 Nanjing was awarded the UNESCO title of "World Capital of Literature". The "World Capital of Literature" Urban Space Programme (2020–2023) is an action plan for the multi-dimensional use of literary resources to empower cities in the international context of the "Global Creative Cities Network". Within the framework of this practical project, this topic will explore how localised design can, on the one hand, empower the local context and, on the other hand, make full use of new opportunities for action that enhance place.

研究生一年级
建筑设计研究（一）：概念设计
周渐佳
课程类型：必修
学时学分：40 学时 / 2 学分

Graduate Program 1st Year
ARCHITECTURAL DESIGN STUDIO 1: CONCEPTUAL DESIGN
· ZHOU Jianjia
Type: Required Course
Study Period and Credits: 40 hours/2 credits

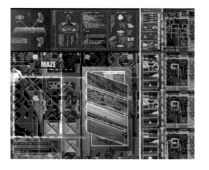

课程内容

疫情使得人们对线上空间的适应与使用突然加速。线上空间曾经被认为是物理空间的附属，却在很短的时间内成为所有活动发生的重要乃至唯一载体。这个过程也将线上空间推到了建筑学科的面前，我们会发现线上空间是与物理空间并行，且有着同等意义的一个新领域，此前极少获得来自学科的关注。这里所说的线上空间不是超越了电影、游戏中作为背景的场景设计，而是突破物理空间的限制，去架构一种新的空间逻辑和交互手段（例如空间的嵌套、循环、时间的重置等），这些成果同样能在物理空间中制造反差，互为验证。为了回应多年来概念设计课程的积累，本次课程以一场展览作为设计对象，以线上空间作为命题的大背景展开。同学们将在为期 8 周的时间内围绕线上展览的空间逻辑、交互手段、相关技术等做全方位的讨论。最终成果同样形成展览，在线上与线下空间同时展示。课程中的所有讲座、小论文、讨论、手稿都将汇编成册，作为对课程的记录。

讲座

交互：凯尔，彼真科技创始人，交互技术专家。
数字生存：陆扬 / 袁松，青年艺术家。
场景：张润泽，TikTok AR Platform。

Course content

The epidemic has led to a sudden acceleration in the adaptation and use of online spaces. Online space was once considered an adjunct to physical space, but it has in a very short period of time become an important occurrence and even the only carrier for all activities to take place. This process has also brought online space to the forefront of the architectural discipline, and we find that online space is a new field that is parallel to physical space and has equal significance, having received very little attention from the discipline before. The online space referred to here goes beyond the design of scenes as backdrops in films and games, but goes beyond the limits of physical space to structure a new spatial logic and means of interaction (e.g. nesting and looping of spaces, resetting of time, etc.), the results of which can also create contrasts and validate each other in the physical space. In response to the accumulation of conceptual design courses over the years, this course takes an exhibition as the design object, with the online space as the broader context for the proposition. The students will discuss all aspects of the spatial logic, means of interaction and related technologies of the online exhibition over a period of 8 weeks. The final result will also be an exhibition, which will be presented in both the online and offline spaces. All the lectures, short papers, discussions, and manuscripts from the course will be compiled into a book as a record of the course.

Lecture

Interaction: Kyle, founder of Pizen Technology, interaction technology expert.
Digital survival: LU Yang/ Yuan Song, young artist.
Scene: Zhang Runze, TikTok AR Platform.

研究生一年级
建筑设计研究（二）：综合设计
程超
课程类型：必修
学时学分：40 学时 / 2 学分

Graduate Program 1st Year
ARCHITECTURAL DESIGN STUDIO 2: COMPREHENSIVE DESIGN · CHENG Chao
Type: Required Course
Study Period and Credits: 40 hours/2 credits

课程内容
建于 1987 年的新图书馆与 1937 年建成的老图书馆以书库进行连接，形成一个以书库为中心向西南两翼发展的布局。2001 年图书馆改造设计在扩大阅览空间的同时，将老图书馆改为校史馆，然而以书库为中心的格局并未改变。仙林校区成为主校区后，鼓楼校区图书馆功能弱化，原书库基本停用。基于鼓楼校园新的发展定位，考虑拆除原图书馆及书库部分，整合现有校史馆功能，规划建设新型的校园学习中心。

基于未来校园发展趋势的学习中心，是集合阅览、自修、研讨、社交、餐饮等功能为一体的多功能综合性建筑，以学生的自主学习为中心，促进校园文化的传播和跨学科的交流融合。它将成为面向未来的学习聚落和知识集市，把学习行为扩展为学习社交。新功能的介入将导致对校园空间的重新诠释，从图书馆到学习中心的转化，是从"以书籍为中心"到"以学生为中心"的转化，建筑的核心功能从各类专业书籍的搜集转化为校园空间网络的交集。

该课程设计以学生熟悉的校园建筑类型为主题，基于校园文脉传承、现状空间格局和未来学习需求，参照校园发展趋势和相关案例，从场所与活动、空间和功能、视线和路径等关系入手，判别设计解决的核心问题，提出明晰的设计策略，强化准确的专业表达，符合相关规范的要求。设计成果要求达到初步设计深度，并选择适合当下建筑产业发展趋势的技术重点进行专项设计。

Course content
The new library, built in 1987, is connected to the old library by a bookstore, forming a layout in which the center of the bookstore develops into two wings to the southwest. With the renovation of the library in 2001, which was designed to expand the reading space and change the function of the old library into the university history museum, the pattern of the bookstore as the center still remained unchanged. After the Xianlin Campus became the main campus, the library function of the Gulou Campus was weakened and the former bookstore was basically decommissioned. Based on the new development orientation of the Gulou Campus, consideration was given to demolishing the former library and bookstore part, integrating the function of the existing University history museum and planning the construction of a new campus learning center.

Based on the future development trend of the campus, the learning center will be a multi-functional and comprehensive building integrating reading, self-study, discussion, social intercourse and catering functions, with students' independent learning as the focus, promoting the dissemination of campus culture and interdisciplinary communication and integration. It will become a future-oriented learning community and knowledge marketplace, extending the act of learning into a learning social. The new functions will lead to a reinterpretation of the campus space, from library to learning center, from "book-centered" to "student-centered". The core function of the building is transformed from a collection of specialist books to a connection of the campus spatial network.

研究生一年级
建筑设计研究（二）：城市设计
华晓宁
课程类型：必修
学时学分：40 学时 / 2 学分

Graduate Program 1st Year
ARCHITECTURAL DESIGN STUDIO 2: URBAN DESIGN · HUA Xiaoning
Type: Required Course
Study Period and Credits: 40 hours/2 credits

课程内容
基础设施是当代城市研究与实践的重要主题。作为社会生产和为居民生活提供公共服务的物质工程设施及其系统，它保障着城市有机体的运行，同时又是城市物质空间系统的重要组成部分，自身便占据了场址，界定了空间，形成了场所，连接成系统，构筑了场域。以往基础设施被仅仅被视作市政工程的专业领域，遵循工具理性，且被传统建筑学忽视多年，许多基础设施已成为城市中消极和被动的要素。然而对于未来都市而言，它已不再仅仅是边缘化、辅助性和服务性的角色，而是一种重要的城市操作性对象与媒介。如何对其"赋能"，将其转化为城市中更为积极、能动的要素，成为激发城市生活的新型"触媒"，是本课题的主要目标。

成果要求
对宁芜铁路秦虹段（东起中和桥道口，西至应天大街高架）及其沿线周边环境进行深入调研，分析存在问题与矛盾，了解沿线居民需求，构想该区段未来愿景，自行拟定任务书，提出改造更新策略，完成方案设计。方案须综合考虑城市、建筑、环境景观，进行整合设计。

Course content
Infrastructure is an important theme in contemporary urban research and practice. As a physical engineering facility and system that provides public services for social production and residential life, it guarantees the functioning of the urban organism and is an important part of the physical-spatial system of the city, occupying the site, defining the space, forming the place, connecting the system and constructing the field. In the past, infrastructure was regarded as a specialised field of civil engineering, following an instrumental rationale, and ignored by traditional architecture for many years. Much infrastructure has become a passive and reactive element of the city. For the city of the future, however, it is no longer just a marginal, auxiliary and service role, but an important urban operable object and medium. The main objective of this project is to "empower" it and transform it into a more active and dynamic element of the city, a new type of "catalyst" that will stimulate urban life.

Achievement requirements
To conduct an in-depth study of the Qinhong section of the Ningwu Railway (from the Zhonghe Bridge crossing in the east to the Yingtian Street elevated in the west) and its surroundings, analyse the problems and contradictions, understand the needs of the residents along the line, conceptualise the future vision of the section, draw up the mission statement, propose a renovation and regeneration strategy and complete the design. The scheme will be designed in an integrated manner, taking into account the urban, architectural and environmental landscape.

研究生一年级
建筑设计研究（二）：城市设计
胡友培
课程类型：必修
学时学分：40 学时 / 2 学分

Graduate Program 1st Year
ARCHITECTURAL DESIGN STUDIO 2: URBAN DESIGN
· HU Youpei
Type: Required Course
Study Period and Credits: 40 hours/2 credits

任务与限定
半城市化是口号与立场，也是任务与限定。为了给设计试验一个刚性的约束，人为将设计范围内可建设用地缩减为原来的50%，而另一半则保持原本的自然属性。用50%的土地实现100%的人口和建设量，并不是简单地将建筑高度翻番，继续保持自然与城市的分立，而是强迫设计者丢掉所有的城市既有模式与成熟的建筑类型学，在设计的底层，展开艰难的创新。50%并不是教条与精确的计算，而更像是一种象征意义上的武断，为无边的想象设定一个可供锚固的起点。

工作与内容
半城市化要求在两个层面对都市区形态学做出全新的尝试。其一是都市区的架构层次，即突破传统的格网新城模式，构想一种新颖的区域空间组织模式，以更加有效地实现自然系统与人工系统的共存，保证都市生活有序展开。其二是与构架匹配的建筑学层次，即突破传统的城市建筑类型，创造、拼贴、裁剪、再加工各种建筑的和人工物的可能空间形式，以从城市物质形态的底层，为都市区架构提供物质性支撑，并激发可能的半城市化的生活形态。

Tasks and limits
Semi-urbanization is the slogan and position, and also the task and limitation. In order to provide a rigid constraint for the design experiment, the constructible land within the design scope is reduced to 50% of the original, while the other half is kept with the original natural properties. To achieve 100% population and construction volume with 50% land is not to simply the double the height of buildings and keep the separation between nature and city, but to force designers to discard all existing urban patterns and mature architectural typologies and carry out difficult innovations at the bottom of the design. Fifty percent is not dogma or precise calculation, but rather symbolic arbitrariness, setting an anchor for the boundless imagination.

Work and content
Semi-urbanization requires a new attempt to the morphology of metropolitan areas at two levels. One is the structure level of metropolitan area. It breaks through the traditional grid new city mode, conceive a new regional spatial organization mode, in order to realize the coexistence of natural system and artificial system more effectively, and ensure the orderly development of urban life. The other one is the architectural level matching with the architecture. It breaks through the traditional urban architectural types, creates, collages, cuts and reprocesses the possible spatial forms of various architectural and artificial objects, so as to provide material support for the metropolitan architecture from the bottom of the urban material form and stimulate the possible semi-urban living form.

研究生国际教学工作坊
An Urban Sharawaggi 城市设计研究
安东尼奥·彼得罗·拉蒂尼，胡友培
课程类型：选修
学时学分：40 学时 / 2 学分

Postgraduate International Design Studio
STUDY ON URBAN DESIGN OF AN URBAN SHARAWAGGI
· Antonio Pietro Latini, HU Youpei
Type: Elective Course
Study Period and Credits: 40 hours/ 2 credits

教学目标
工作坊试图探索如何使普通人的生活场所比现在更加愉悦和可持续，重点是环境形式和美学品质。为了克服当代城市令人压抑的单调性和混乱状况，我们依靠中国传统的园林提供的刺激：一种在和谐多样的环境中结合多种环境设计元素的艺术。本次课程主要涵盖了诸如街道/街区格局、图底关系、道路和开放空间类型设计、地块、建筑类型、建筑密度、流线设计等设计主题。

教学内容
课程前期学生能够积累基本的城市设计学科背景知识，且每周进行专题设计。然后在此基础上，由三人一组合作进行城市设计。选址位于南京江宁区东流村，基地面积约300 hm²。前期的图形练习在最后的小组设计项目中达到高潮，在这个过程中把空间构成、景观设计和规划设计在创新的美学研究中结合起来。

Teaching objectives
The workshop tries to explore how to make ordinary people's living place more pleasant and sustainable than now, focusing on the environmental form and aesthetic quality. In order to overcome the depressing monotony and chaos of contemporary cities, we rely on the stimulation provided by Chinese traditional gardens: an art that combines a variety of environmental design elements in a harmonious and diverse environment. This course mainly covers design topics such as the street / block pattern, figure and background relationship, road and open space type design, plot, building type, building density, streamline design and so on.

Teaching content
In the early stage of the course, students can accumulate basic background knowledge of urban design and carry out special design every week. Then, on this basis, three people work together to carry out urban design. The site is located in Dongliu Village, Jiangning District, Nanjing, covering an area of about 300 hm². The early graphic practice reaches a climax in the final group design project. In this process, the space composition, landscape design and planning design are combined in the innovative aesthetic research.

研究生国际教学工作坊
电影建筑工作坊
弗朗索瓦·潘兹 鲁安东
课程类型：选修
学时学分：18 学时 / 1 学分

Postgraduate International Design Studio
CINEMATIC ARCHITECTURE WORKSHOP · Francois Penz, LU Andong
Type: Elective Course
Study Period and Credits: 18 hours/1 credits

课程内容

受 2021 年度南京大学国际合作与交流处"引进人文社科类资深海外专家重点支持计划"支持，剑桥大学弗朗索瓦·潘兹教授开设线上研究生课程"电影建筑"，本专业选课学生 26 人。

课程包括学术讲座系列和影像拍摄工作坊两个板块：学术讲座系列共 8 讲，包括 5 次内部讲座及 3 次线上直播讲座，听课总人数超过 9 600 人次；影像拍摄工作坊进一步扩大参加范围，面向南京大学校内其他学科进行了招募。工作坊时长 7 d，由学生 4 人一组，以南京为研究对象，制作 90 s 短片呈现了后疫情时代公共空间的四个维度。拍摄成果进行了线上评图，参加人数超过 20 000 人。此外，成果短片受邀在 2021 年第 27 届世界建筑师大会放映，受到主办方的高度赞誉。

Course content

Supported by the "Key Support Program for the Introduction of Senior Overseas Experts in Humanities and Social Sciences", a key university-level intellectual attraction project in 2021, Professor Francois Penz from the University of Cambridge offered the online graduate course "Cinematic Architecture", which was attended by 26 students.
The course consists of two sections: academic lecture series and video shooting workshop. The academic lecture series consists of eight lectures, including five internal lectures and three online live lectures, with a total attendance of more than 9 600; the video shooting workshop further expanded the scope of participation and was open to other disciplines within Nanjing University. The workshop lasted 7 days and was conducted in groups of 4 students, with Nanjing as the research object. The results of the filming were evaluated online with over 20 000 participants. In addition, the resulting short film was invited to be screened at the 27th World Congress of Architects in 2021 and received high praise from the organizers.

研究生国际教学工作坊
设计研究学导论
默里·弗雷泽 鲁安东
课程类型：选修
学时学分：18 学时 / 1 学分

Postgraduate International Design Studio
INTRODUCTION TO DESIGN RESEARCH · Murray Fraser, LU Andong
Type: Elective Course
Study Period and Credits: 18 hours/1 credits

课程内容

受 2021 年度校级重点引智项目"引进人文社科类资深海外专家重点支持计划"支持，伦敦大学学院默里·弗雷泽教授开设线上研究生课程"设计研究学导论"，课程聚焦当代设计研究的基本问题和前沿思想，将理论研究与设计实践对比印证，本专业选课学生共 26 人。

课程包括学术讲座系列、线上工作坊、文献精读 3 个板块。

1. 学术讲座系列共 3 讲，仅限选课学生参加。
2. 线上工作坊共进行了 4 次线上研讨，参加教学的嘉宾包括多位英国皇家建筑师学会安妮·斯宾克建筑教育终身成就奖得主、诸多国内外知名建筑师和理论家。线上直播听课总人数超过 30 000 人次。
3. 配合课程开展文献精读计划，在全球范围进行招募并选拔出 8 名青年学人，与本校学生合作完成翻译任务。围绕翻译成果进行了 2 次汇报研讨会。

Course content

Supported by the "Key Support Program for the Introduction of Senior Overseas Experts in Humanities and Social Sciences", a key university-level intellectual attraction project in 2021, Professor Murray Fraser of University College London offered the online graduate course "Introduction to Design Research", which focuses on the fundamental issues and cutting-edge ideas of contemporary design research, and compares theoretical research with design practice, which was attended by 26 students.
The course included three sections: academic lecture series, online workshops, and intensive reading of literature:
1. The academic lecture series consisted of three lectures, which were restricted to students taking the course.
2. The online workshop series consisted of four online seminars with the participation of several RIBA Annie Spink Lifetime Achievement Award winners in architectural education and many famous architects and theorists from home and abroad. The total number of attendees of the live online workshops exceeded 30 000.
3. In conjunction with the course, a literature intensive reading program was conducted, and eight young scholars were recruited and selected from around the world to work with our students to complete the translation tasks. Two reporting seminars were held on the results of the translations.

研究生国际教学工作坊

建构共生与未来环境建造——国际之声工作坊

凯瑞·希瑞斯　丁沃沃

课程类型：选修

学时学分：18 学时 / 1 学分

Postgraduate International Design Studio
CONSTRUCTING COEXISTENCE AND FUTURES OF ENVIRONMENT-MAKING: INTERNATIONAL VOICES WORKSHOP · Cary Siress, DING Wowo
Type: Elective Course
Study Period and Credits: 18 hours/1 credits

课程内容

　　2020 年对于世界各国来说是极为不平凡的一年，蔓延到全世界且至今尚未结束的疫情证实了人们在世纪之初的预感，世界正在巨变。此时，南京大学建筑学迎来了办学 20 周年。20 年的历程不长，但充满了探索，面对巨变的世界，南大建筑将肩负责任，重新踏上新的探索征程。建筑学在探索，建筑学需要探索，建筑学需要在探索中重构。这是一个全新的征途，目标和行动都充满不确定性和挑战，也充满期待。为此，我们搭建了国际前沿系列讲座的平台，邀请国际一流大学中对此有思考的学者，尤其是有思想的年轻的学者，请其针对人类共同的生存环境的问题阐述自己的观点，在论述中放出思想的火花。

Course content

The year 2020 is an extraordinary year for all countries in the world. The epidemic that has spread all over the world and has not yet ended has confirmed people's hunch at the beginning of the century that the world is changing dramatically. At this time, Nanjing University Architecture ushered in its 20th anniversary. The 20-year course is not long, but it is full of exploration. In the face of a world of great changes, NJU architecture will shoulder the responsibility and embark on a new journey of exploration. Architecture is exploring, architecture needs to be explored, and architecture needs to be reconstructed in exploration. This is a new journey, with goals and actions full of uncertainties, challenges and expectations. To this end, we have set up a platform for a series of international cutting-edge lectures, and invited thoughtful scholars from world-class universities, especially thoughtful young scholars, to elaborate their views on the problems of the common living environment of mankind, and to spark their thoughts in the discussion.

建筑理论课程
ARCHITECTURAL THEORY COURSES

本科二年级
建筑导论・赵辰　等
课程类型：必修
学时 / 学分：36 学时 /2 学分

Undergraduate Program 2nd Year
INTRODUCTOTY GUIDE TO ARCHITECTURE
• ZHAO Chen, et al.
Type: Required Course
Study Period and Credits:36 hours / 2 credits

课程内容
1. 建筑学的基本定义
　第一讲： 建筑与设计 / 赵辰
　第二讲： 建筑与城市 / 丁沃沃
　第三讲： 建筑与生活 / 张雷
2. 建筑的基本构成
　（1）建筑的物质构成
　第四讲： 建筑的物质环境 / 赵辰
　第五讲： 建筑与节能技术 / 郜志
　第六讲： 建筑与生态环境 / 吴蔚
　第七讲： 建筑的环境智慧 / 窦平平
　（2）建筑的文化构成
　第八讲： 建筑与人文、艺术、审美 / 赵辰
　第九讲： 建筑与环境景观 / 华晓宁
　第十讲： 中西方风景观念与设计 / 史文娟
　第十一讲： 建筑与身体经验 / 鲁安东
　（3）建筑师职业与建筑学术
　第十二讲： 建筑与表现 / 赵辰
　第十三讲： 建筑与乡村复兴 / 黄华青
　第十四讲： 建筑与数字技术 / 钟华颖
　第十五讲： 建筑师的职业技能与社会责任 / 傅筱

Course Content
1. Basic definition of architecture
Lecture 1: Architecture and design / ZHAO Chen
Lecture 2: Architecture and urbanization / DING Wowo
Lecture 3: Architecture and life / ZHANG Lei
2. Basic attribute of architecture
(1) Physical attribute
Lecture 4: Physical environment of architecture / ZHAO Chen
Lecture 5: Architecture and energy saving / GAO Zhi
Lecture 6: Architecture and ecological environment / WU Wei
Lecture 7: Environmental intelligence in architecture / DOU Pingping
(2) Cultural attribute
Lecture 8: Architecture and civilization, arts, aesthetic / ZHAO Chen
Lecture 9: Architecture and landscaping environment / HUA Xiaoning
Lecture 10: Landscaping View and Design in Comparison of China and West/SHI Wenjuan
Lecture 11: Architecture and body / LU Andong
(3) Architect: profession and academy
Lecture 12: Architecture and presentation / ZHAO Chen
Lecture 13: Architecture and rural rival / HUANG Huaqing
Lecture 14: Architectural and digital technology / ZHONG Huaying
Lecture 15: Architect's professional technique and social responsibility / FU Xiao

---

本科三年级
建筑设计基本原理・周凌
课程类型：必修
学时 / 学分：36 学时 /2 学分

Undergraduate Program 3rd Year
BASIC THEORY OF ARCHITECTURAL DESIGN
• ZHOU Ling
Type: Required Course
Study Period and Credits:36 hours / 2 credits

教学目标
　本课程是建筑学专业本科生的专业基础理论课程。本课程的任务主要是介绍建筑设计中形式与类型的基本原理。形式原理包含历史上各个时期的设计原则，类型原理讨论不同类型建筑的设计原理。
课程要求
　1. 讲授大纲的重点内容；
　2. 通过分析实例启迪学生的思维，加深学生对有关理论及其应用、工程实例等内容的理解；
　3. 通过对实例的讨论，引导学生运用所学的专业理论知识，分析、解决实际问题。
课程内容
　1. 形式与类型概述
　2. 古典建筑形式语言
　3. 现代建筑形式语言
　4. 当代建筑形式语言
　5. 类型设计
　6. 材料与建造
　7. 技术与规范
　8. 课程总结

Teaching objectives
This course is a basic theory course for the undergraduate students of architecture. The main purpose of this course is to introduce the basic principles of the form and type in architectural design. Form theory contains design principles in various periods of history; type theory discusses the design principles of different types of the building.
Course requirement
1. Teach the key elements of the outline;
2. Enlighten students' thinking and enhance students' understanding of the theories, its applications and project examples through analyzing examples;
3. Help students to use the professional knowledge to analysis and solve practical problems through the discussion of examples.
Course content
1. Overview of forms and types
2. Classical architecture form language
3. Modern architecture form language
4. Contemporary architecture form language
5. Type design
6. Materials and construction
7. Technology and specification
8. Course summary

本科三年级
居住建筑设计与居住区规划原理・冷天 刘铨
课程类型：必修
学时 / 学分：36 学时 /2 学分

Undergraduate Program 3rd Year
THEORY OF HOUSING DESIGN AND RESIDENTIAL PLANNING • LENG Tian, LIU Quan
Type: Required Course
Study Period and Credits:36 hours / 2 credits

课程内容
　第一讲： 课程概述
　第二讲： 居住建筑的演变
　第三讲： 套型空间的设计
　第四讲： 套型空间的组合与单体设计（一）
　第五讲： 套型空间的组合与单体设计（二）
　第六讲： 居住建筑的结构、设备与施工
　第七讲： 专题讲座：住宅的适应性，支撑体住宅
　第八讲： 城市规划理论概述
　第九讲： 现代居住区规划的发展历程
　第十讲： 居住区的空间组织
　第十一讲： 居住区的道路交通系统规划与设计
　第十二讲： 居住区的绿地景观系统规划与设计
　第十三讲： 居住区公共设施规划、竖向设计与管线综合
　第十四讲： 专题讲座：住宅产品开发
　第十五讲： 专题讲座：住宅产品设计实践
　第十六讲： 课程总结，考试答疑

Course content
Lecture 1: Introduction of the course
Lecture 2: Development of the residential building
Lecture 3: Design of dwelling space
Lecture 4: Dwelling space arrangement and monomer building design (1)
Lecture 5: Dwelling space arrangement and monomer building design (2)
Lecture 6: Structure, facility and construction of residential buildings
Lecture 7: Adaptability of residential building, supporting house
Lecture 8: Introduction of the theories of urban planning
Lecture 9: History of modern residential planning
Lecture 10: Organization of residential space
Lecture 11: Traffic system planning and design of residential area
Lecture 12: Landscape planning and design of residential area
Lecture 13: Public facilities and infrastructure system
Lecture 14: Real estate development
Lecture 15: The practice of residential planning and housing design
Lecture 16: Summary, question of the test

研究生一年级
现代建筑设计基础理论 · 周凌，梁宇舒
课程类型：必修
学时/学分：18学时/1学分

Graduate Program 1st Year
PRELIMINARIES IN MODERN ARCHITECTURAL DESIGN · ZHOU Lin, LIANG Yushu
Type: Required Course
Study Period and Credits: 18 hours/1 credit

教学目标
建筑可以被抽象到最基本的空间围合状态来面对它所必须解决的基本的适用问题，用最合理、最直接的空间组织和建造方式去解决问题，以普通材料和通用方法去回应复杂的使用要求，是建筑设计所应该关注的基本原则。
课程要求
1. 讲授大纲的重点内容；
2. 通过分析实例启迪学生的思维，加深学生对有关理论及其应用、工程实例等内容的理解；
3. 通过对实例的讨论，引导学生运用所学的专业理论知识，分析、解决实际问题。
课程内容
1. 基本建筑的思想
2. 基本空间的组织
3. 建筑类型的抽象与还原
4. 材料的运用与建造问题
5. 场所的形成及其意义
6. 建筑构思与设计概念

Teaching objectives
The architecture can be abstracted into spatial enclosure state to encounter basic application problems which must be settled. Solving problems with most reasonable and direct spatial organization and construction mode, and responding to operating requirements with common materials and general methods are basic principle concerned by building design.
Course requirement
1. To teach key contents of syllabus;
2. To inspire students' thinking, deepen students' understanding on such contents as relevant theories and their application and engineering examples through case analysis.
3. To help students to use professional theories to analyze and solve practical problems through discussion of instances.
Course content
1. Basic architectural thought
2. Basic spacial organization
3. Abstraction and restoration of architectural types
4. Utilization and construction of materials
5. Formation of site and its meaning
6. Architectural conception and design concept

研究生一年级
建筑与规划研究方法 · 鲁安东 等
课程类型：必修
学时/学分：18学时/1学分

Graduate Program 1st Year
RESEARCH METHOD OF ARCHITECTURE AND URBAN PLANING
· LU Andong, et al.
Type: Required Course
Study Period and Credits: 18 hours/1 credit

教学目标
面向学术型硕士研究生的必修课程。它将向学生全面地介绍学术研究的特性、思维方式、常见方法以及开展学术研究必要的工作方式和写作规范。考虑到不同领域研究方法的差异，本课程的授课和作业将以专题的形式进行组织，包括建筑研究概论、设计研究、科学研究、历史理论研究4个模块。学生通过各模块的学习可以较为全面地了解建筑学科内主要的研究领域及相应的思维方式和研究方法。
课程要求
将介绍建筑学科的主要研究领域和当代研究前沿，介绍"研究"的特性、思维方式、主要任务、研究的工作架构以及什么是好的研究，帮助学生建立对"研究"的基本认识；介绍文献检索和文献综述的规范和方法；介绍常见的定量研究、定性研究和设计研究的工作方法以及相应的写作规范。
课程内容
1. 综述
2. 文献
3. 科学研究及其方法
4. 科学研究及其写作规范
5. 历史理论研究及其方法
6. 历史理论研究及其写作规范
7. 设计研究及其方法
8. 城市规划理论概述

Teaching objectives
It is a compulsory course to MA. It comprehensively introduces features, ways of thinking and common methods of academic research, and necessary manners of working and writing standard for launching academic research to students. Considering differences of research methods among different fields, teaching and assignment of the course will be organized in the form of special topic, including four parts: introduction to architectural study, design study, scientific study and historical theory study. Through the study of all parts, students can comprehensively understand main research fields and corresponding ways of thinking and research methods of architecture.
Course requirement
The course introduces main research fields and contemporary research frontier of architecture, features, ways of thinking and main tasks of "research", working structure of research, and definition of good research to help students form basic understanding of "research". The course also introduces standards and methods of literature retrieval and review, and working methods of common quantitative research, qualitative research and design research, and their corresponding writing standards.
Course content
1. Review
2. Literature
3. Scientific research and methods
4. Scientific research and writing standards
5. Historical theory study and methods
6. Historical theory study and writing standards
7. Design research and methods
8. Overview of urban planning theory

城市理论课程
URBAN THEORY COURSES

本科四年级
城市设计及其理论 · 胡友培
课程类型:必修
学时/学分: 36学时/2学分

Undergraduate Program 4th Year
URBAN DESIGN AND THEORY · HU Youpei
Type: Required Course
Study Period and Credits: 36 hours / 2 credits

课程内容
第一讲:课程概述
第二讲:城市设计技术术语——城市规划相关术语;城市形态相关术语;城市交通相关术语;消防相关术语
第三讲:城市设计方法——文本分析:城市设计上位规划;城市设计相关文献;文献分析方法
第四讲:城市设计方法——数据分析:人口数据分析与配置;交通流量数据分析;功能分配数据分析;视线与高度数据分析;城市空间数据模型的建构
第五讲:城市设计方法——城市肌理分类;城市肌理分类概述;肌理形态与建筑容量;肌理形态与开放空间;肌理形态与交通流量;城市绿地指标体系
第六讲:城市设计方法——城市路网组织;城市道路结构与交通结构概述;城市路网与城市功能;城市路网与城市空间;城市路网与市政设施;城市道路断面设计
第七讲:城市设计方法——城市设计表现:城市设计分析图;城市设计概念表达;城市设计成果解析图;城市设计地块深化设计表达;城市设计空间表达
第八讲:城市设计的历史与理论——城市设计的历史意义;城市设计理论的内涵
第九讲:城市路网形态——路网形态的类型和结构;路网形态与肌理;路网形态的变迁
第十讲:城市空间——城市空间的类型;城市空间结构;城市空间形态;城市空间形态的变迁
第十一讲:城市形态学——英国学派;意大利学派;法国学派;空间句法
第十二讲:城市形态的物理环境——城市形态与物理环境;城市形态与环境研究;城市形态与环境测评;城市形态与环境操作
第十三讲:景观都市主义——景观都市主义的理论、操作与范例
第十四讲:城市自组织现象及其研究——城市自组织现象的魅力与问题;城市自组织系统研究方法;典型自组织现象案例研究
第十五讲:建筑学图式理论与方法——图式理论的研究;建筑学图式的概念;图式理论的应用;作为设计工具的图式;当代城市语境中的建筑学图式理论探索
第十六讲:课程总结

Course content
Lecture 1: Introduction
Lecture 2: Technical terms—terms of urban planning, urban morphology, urban traffic and fire protection
Lecture 3: Urban design methods—document analysis: urban planning and policies; relative documents; document analysis techniques and skills
Lecture 4: Urban design methods—data analysis: data analysis of demography, traffic flow, function distribution, visual and building height; modelling urban spatial data
Lecture 5: Urban design methods—classification of urban fabrics: introduction of urban fabrics; urban fabrics and floor area ratio; urban fabrics and open space; urban fabrics and traffic flow; criteria system of urban green space
Lecture 6: Urban design methods—organization of urban road network: introduction; urban road network and urban function; urban road network and urban space; urban road network and civic facilities; design of urban road section
Lecture 7: Urban design methods—representation skills of urban design: mapping and analysis; conceptual diagram; analytical representation of urban design; representation of detail design; spatial representation of urban design
Lecture 8: Brief history and theories of urban design—historical meaning of urban design; connotation of urban design theories
Lecture 9: Form of urban road network—typology, structure and evolution of road network; road network and urban fabrics; changes of road network form
Lecture 10: Urban space—typology, structure, morphology and evolution of urban space
Lecture 11: Urban morphology—Cozen School; Italian School; French School; Space Syntax Theory
Lecture 12: Physical environment of urban forms—urban forms and physical environment; environmental study; environmental evaluation and environmental operations
Lecture 13: Landscape urbanism—ideas, theories, operations and examples of landscape urbanism
Lecture 14: Researches on the phenomena of the urban self-organization—charms and problems of urban self-organization phenomena; research methodology on urban self-organization phenomena; case studies of urban self-organization phenomena
Lecture 15: Theory and method of architectural diagram—theoretical study on diagrams; concepts of architectural diagrams; application of diagram theory; diagrams as design tools; theoretical research of architectural diagrams in contemporary urban context
Lecture 16: Summary

本科四年级
景观规划设计及其理论 · 尹航
课程类型:选修
学时/学分: 36学时/2学分

Undergraduate Program 4th Year
LANDSCAPE PALNNING DESIGN AND THEORY
· YIN Hang
Type: Elective Course
Study Period and Credits: 36 hours / 2 credits

课程介绍
景观规划设计的对象包括所有的室外环境,景观与建筑的关系往往是紧密而互相影响的,这种关系在城市中表现得尤为明显。景观规划设计及其理论课程希望从景观设计理念、场地设计技术和建筑周边环境塑造等方面开展课程的教学,为建筑学本科生建立更加全面的景观知识体系,并且完善建筑学本科生在建筑场地设计、总平面规划与城市设计等方面的设计能力。
本课程主要从三个方面展开。一是理念与历史:以历史的视角介绍景观学科的发展过程,让学生对景观学科有一个宏观的了解,初步理解景观设计理念的发展;二是场地与文脉:通过阐述景观规划设计与周边自然环境、地理位置、历史文脉和方案可持续性的关系,建立场地与文脉的设计思维;三是景观与建筑:通过设计方法授课、先例分析作业等方式让学生增强建筑的环境意识,了解建筑的场地设计的影响因素、一般步骤与方法,并通过与"建筑设计六"和"建筑设计七"的设计任务书相配合的同步课程设计训练来加强学生景观规划设计的能力。

Course description
The object of landscape planning design includes all outdoor environments; the relationship between landscapes and buildings is often close and interactive, which is especially obvious in a city. This course expects to carry out teaching from perspective of landscape design concept, site design technology, building's peripheral environment creation, etc., to establish a more comprehensive landscape knowledge system for the undergraduate students of architecture, and perfect their design ability in building site design, plane planning and urban design and so on.
This course includes three aspects. First, concept and history: introduce the development process of landscape discipline from a historical perspective, so that students can have a macro understanding of landscape discipline and preliminarily understand the development of landscape design concept; The second is the site and context: the design thinking of the site and context is established by explaining the relationship between the landscape planning and design and the surrounding natural environment, geographical location, historical context and program sustainability; Third, landscape and architecture: through design method teaching and precedent analysis, students can enhance their environmental awareness of architecture, understand the influencing factors, general steps and design methods of building site design, and strengthen their ability of landscape planning and design through synchronous course design training in conjunction with the design tasks of "architectural design VI" and "architectural design VII".

研究生一年级
城市形态与设计方法论 · 丁沃沃
课程类型：必修
学时 / 学分：36 学时 / 2 学分

Graduate Program 1st Year
URBAN MORPHOLOGY AND DESIGN METHOLOGY
• DING Wowo
Type: Required Course
Study Period and Credits: 36 hours / 2 credits

课程介绍
　　建筑学核心理论包括建筑学的认识论和设计方法论两大部分。建筑设计方法论主要探讨设计的认知规律、形式的逻辑、形式语言类型，以及人的行为、环境特征和建筑材料等客观规律对形式语言的选择及形式逻辑的构成策略。为此，设立了以提升建筑设计方法为目的的关于设计方法论的理论课程，作为建筑设计及其理论硕士学位的核心课程。
课程要求
　　1. 理解随着社会转型，城市建筑的基本概念在建筑学核心理论中的地位以及认知的视角。
　　2. 通过理论的研读和案例分析理解建筑形式语言的成因和逻辑，并厘清中、西不同的发展脉络。
　　3. 通过研究案例的解析理解建筑形式语言的操作并掌握设计研究的方法。
课程内容
　　第一讲：序言
　　第二讲：西方建筑学的基础
　　第三讲：中国——建筑的意义
　　第四讲：背景与文献研讨
　　第五讲：历史观与现代性
　　第六讲：现代城市形态演变与解析
　　第七讲：现代城市的"乌托邦"
　　第八讲：现代建筑的意义
　　第九讲：建筑形式的反思与探索
　　第十讲：建筑的量产与城市问题
　　第十一讲："乌托邦"的实践与反思
　　第十二讲：都市实践探索的理论价值
　　第十三讲：城市形态的研究
　　第十四讲：城市空间形态研究的方法
　　第十五讲：回归理性——建筑学方法论的新进展
　　第十六讲：建筑学与设计研究的意义
　　第十七讲：结语与研讨（一）
　　第十八讲：结语与研讨（二）

Course description
Core theory of architecture includes epistemology and design methodology of architecture. Architectural design methodology mainly discusses cognitive laws of design, logic of forms and types of formal language, and the choice of formal language from objective laws such as human behaviors, environmental features and building materials, and composition strategy of formal logic. Thus, the theory course about design methodology to promote architectural design methods is established as the core course of orientations of architectural design and theory master degree.
Course requirement
1. To understand the status and cognitive perspective of basic concept of urban buildings in the core theory of architecture with the social transformation.
2. To understand the reason and logic of architectural formal language and different development process in China and the West through theory reading and case analysis.
3. To understand the operation of architectural formal language and grasp methods of design study by analyzing cases.
Course content
Lecture 1: Introduction
Lecture 2: Foundation of western architecture
Lecture 3: China: meaning of architecture
Lecture 4: Background and literature discussion
Lecture 5: Historicism and modernity
Lecture 6: Analysis and morphological evolution of modern city
Lecture 7: "Utopia" of modern city
Lecture 8: Meaning of modern architecture
Lecture 9: Reflection and exploration of architectural forms
Lecture 10: Mass production of buildings and urban problems
Lecture 11: Practice and reflection of "Utopia"
Lecture 12: Theoretical value of exploration on urban practice
Lecture 13: Study on urban morphology
Lecture 14: Methods of urban spatial morphology study
Lecture 15: Return to rationality: new developments of methodology on architecture
Lecture 16: Meaning of architecture and design study
Lecture 17: Conclusion and discussion (1)
Lecture 18: Conclusion and discussion (2)

研究生一年级
景观都市主义理论与方法 · 华晓宁
课程类型：选修
学时 / 学分：18 学时 / 1 学分

Graduate Program 1st Year
THEORY AND METHOD OF LANDSCAPE URBANISM
• HUA Xiaoning
Type: Elective Course
Study Period and Credits: 18 hours / 1 credit

课程介绍
　　本课程作为国内首次以景观都市主义相关理论与策略为教学内容的尝试，介绍了景观都市主义思想产生的背景、缘起及其主要理论观点，并结合实例，重点分析了其在不同的场址和任务导向下发展起来的多样化的实践策略和操作性工具。
课程要求
　　1. 要求学生了解景观都市主义思想产生的背景、缘起和主要理念。
　　2. 要求学生能够初步运用景观都市主义的理念和方法分析和解决城市设计问题，从而在未来的城市设计实践中强化景观整合意识。
课程内容
　　第一讲：从图像到效能——景观都市实践的历史演进与当代视野
　　第二讲：生态效能导向的景观都市实践（一）
　　第三讲：生态效能导向的景观都市实践（二）
　　第四讲：社会效能导向的景观都市实践
　　第五讲：基础设施景观都市实践
　　第六讲：当代高密度城市中的地形学
　　第七讲：城市图绘与图解
　　第八讲：从原形到系统——AA 景观都市主义

Course description
Combining relevant theories and strategies of landscape urbanism firstly in China, the course introduces the background, origin and main theoretical viewpoint of landscape urbanism, and focuses on diversified practical strategies and operational tools developed under different orientations of site and task with examples.
Course requirement
1. Students are required to understand the background, origin and main concept of landscape urbanism.
2. Students are required to preliminarily utilize the concept and method of landscape urbanism to analyze and solve the problem of urban design, so as to strengthen landscape integration consciousness in the future.
Course content
Lecture 1: From pattern to performance: historical revolution and contemporary view of practice of landscape urbanism
Lecture 2: Eco-efficiency-oriented practice of landscape urbanism (1)
Lecture 3: Eco-efficiency-oriented practice of landscape urbanism (2)
Lecture 4: Social efficiency-oriented practice of landscape urbanism
Lecture 5: Infrastructure practice in landscape urbanism
Lecture 6: Geomorphology in contemporary high-density cities
Lecture 7: Urban painting and diagrammatizing
Lecture 8: From prototype to system—AA landscape urbanism

历史理论课程
HISTORY THEORY COURSES

本科二年级
中国建筑史（古代）• 赵辰 史文娟 赵潇欣
课程类型：必修
学时 / 学分：36 学时 /2 学分

Undergraduate Program 2nd Year
HISTORY OF CHINESE ARCHITECTURE (ANCIENT)
• ZHAO Chen, SHI Wenjuan, ZHAO Xiaoxin
Type: Required Course
Study Period and Credits: 36 hours / 2 credits

教学目标
本课程作为本科建筑学专业的历史与理论课程，目标在于培养学生的史学研究素养与对中国建筑及其历史的认识两个层面。在史学理论上，引导学生理解建筑史学这一交叉学科的多种视角，并从多种相关学科层面对学生进行基本史学研究方法的训练与指导。中国建筑史层面，帮助学生建立对中国传统建筑的营造特征与文化背景的构架性的认识体系。

课程内容
中国建筑史学七讲与方法论专题。七讲总体走向从微观到宏观，整体以建筑单体—建筑群体—聚落与城市—历史地理为序；从物质性到文化，建造技术—建造制度—建筑的日常性—纪念性—政治与宗教背景—美学追求。方法论专题包括建筑考古学、建筑技术史、人类学、美术史等层面。

Teaching objectives
As a mandatory historical & theoretical course for undergraduate students, this course aims at two aspects of training: the basic academic capability of historical research and the understanding of Chinese architectural history. It will help students to establish a knowledge frame, that the discipline of History of Architecture as a cross-discipline, is supported and enriched by multiple neighboring disciplines and that the features and development of Chinese Architecture roots deeply in the natural and cultural background.
Course content
The course composes seven lectures on Chinese Architecture and a series of lectures on methodology. The seven lectures follow a route from individual to complex, from physical building to the intangible technique and to the cultural background, from technology to institution, to political and religious background, and finally to aesthetic pursuit. The special topics on methodology include building archaeology, history of building science and technology, anthropology, art history and so on.

本科二年级
外国建筑史（古代）• 王骏阳
课程类型：必修
学时 / 学分：36 学时 /2 学分

Undergraduate Program 2nd Year
HISTORY OF WESTERN ARCHITECTURE (ANCIENT)
• WANG Junyang
Type: Required Course
Study Period and Credits: 36 hours / 2 credits

教学目标
本课程力图对西方建筑史的脉络做一个整体勾勒，使学生在掌握重要的建筑史知识点的同时，对西方建筑史在2000多年里的变迁的结构转折（不同风格的演变）有深入的理解。本课程希望学生对建筑史的发展与人类文明发展之间的密切关联有所认识。

课程内容
1. 概论 2. 希腊建筑 3. 罗马建筑 4. 中世纪建筑
5. 意大利的中世纪建筑 6. 文艺复兴 7. 巴洛克
8. 美国城市 9. 北欧浪漫主义 10. 加泰罗尼亚建筑
11. 先锋派 12. 德意志制造联盟与包豪斯
13. 苏维埃的建筑与城市 14. 1960年代的建筑
15. 1970年代的建筑 16. 答疑

Teaching objectives
This course seeks to give an overall outline of Western architectural history, so that enable students to master important knowledge points of architectural history, and may have an in-depth understanding of the structural transition (different styles of evolution) of Western architectural history in the past 2000 years. This course hopes that students can understand the close association between the development of architectural history and the development of human civilization.
Course content
1. Generality 2. Greek architecture 3. Roman architecture
4. The Middle Ages architecture
5. The Middle Ages architecture in italy 6. Renaissance
7. Baroque 8. American cities 9. Nordic romanticism
10. Catalonian architectures 11. Avant-garde
12. German manufacturing alliance and Bauhaus
13. Soviet architecture and cities 14. 1960's architecture
15. 1970's architectures 16. Answer questions

本科三年级
外国建筑史（当代）• 胡恒
课程类型：必修
学时 / 学分：36 学时 /2 学分

Undergraduate Program 3rd Year
HISTORY OF WESTERN ARCHITECTURE (MODERN)
• HU Heng
Type: Required Course
Study Period and Credits: 36 hours / 2 credits

教学目标
本课程力图用专题的方式对文艺复兴时期的7位代表性的建筑师与5位当代的重要建筑师作品做一细致的讲解。本课程将重要建筑师的全部作品尽可能在课程中梳理一遍，使学生能够全面掌握重要建筑师的设计思想、理论主旨、与时代的特殊关联、在建筑史中的意义。

课程内容
1. 伯鲁乃列斯基 2. 阿尔伯蒂 3. 伯拉孟特
4. 米开朗琪罗（1）5. 米开朗琪罗（2）6. 罗马诺
7. 桑索维诺 8. 帕拉蒂奥（1）9. 帕拉蒂奥（2）
10. 赖特 11. 密斯 12. 勒·柯布西耶（1）
13. 勒·柯布西耶（2）14. 海杜克 15. 妹岛和世
16. 答疑

Teaching objectives
This course seeks to make a detailed explanation to the works of 7 representative architects in the Renaissance period and 5 important modern and contemporary architects in a special way. This course will try to reorganize all works of these important architects, so that the students can fully grasp their design ideas, theoretical subject and their particular relevance with the era and the significance in the architectural history.
Course content
1. Brunelleschi 2. Alberti 3. Bramante
4. Michelangelo(1) 5. Michelangelo(2)
6. Romano 7. Sansovino 8. Palladio(1) 9. Palladio(2)
10. Wright 11. Mies 12. Le Corbusier(1) 13. Le Corbusier(2)
14. Hejduk 15. Kazuyo Sejima
16. Answer questions

本科三年级
中国建筑史（近现代）• 赵辰 冷天
课程类型：必修
学时 / 学分：36 学时 /2 学分

Undergraduate Program 3rd Year
HISTORY OF CHINESE ARCHITECTURE (MODERN)
• ZHAO Chen, LENG Tian
Type: Required Course
Study Period and Credits: 36 hours / 2 credits

课程介绍
本课程作为本科建筑学专业的历史与理论课程，是中国建筑史教学中的一部分。在中国与西方的古代建筑历史课程的基础上，了解中国社会进入近代，以至于现当代的发展进程。
在对比中西方建筑文化的基础之上，建立对中国近现代建筑的整体认识。深刻理解中国传统建筑文化在近代以来与西方建筑文化的冲突与相融之下，逐步演变发展至今天成为世界建筑文化的一部分之意义。

Course description
As the history and theory course for undergraduate students of Architecture, this course is part of the teaching of History of Chinese Architecture. Based on the earlier studying of Chinese and Western history of ancient architecture, understand the evolution progress of Chinese society's entry into modern times and even contemporary age.
Based on the comparison between Chinese and western building culture, establish the overall understanding of China's modern and contemporary buildings. Have further understanding of the significance of China's traditional building culture's gradual evolution into one part of today's world building culture under the conflict and blending with Western building culture in modern times.

研究生一年级
建筑理论研究 · 赵辰
课程类型：必修
学时/学分：18学时/1学分

Graduate Program 1st Year
STUDIES OF ARCHITECTURAL THEORY · ZHAO Chen
Type: Required Course
Study Period and Credits: 18 hours / 1 credit

课程介绍
了解中、西方学者对中国建筑文化诠释的发展过程，理解新的建筑理论体系中对中国筑重新诠释的必要性，学习重新诠释中国建筑文化的建筑观念与方法。

课程内容
1. 本课的总览和基础
2. 中国建筑：西方人的诠释与西方建筑观念的改变
3. 中国建筑：中国人的诠释以及中国建筑学术体系的建立
4. 木结构体系：中国建构文化的意义
5. 住宅与园林：中国人居文化意义
6. 宇宙观的和谐：中国城市文化的意义
7. 讨论

Course description
Understand the development process of Chinese and western scholars' interpretation of Chinese architectural culture, understand the necessity of reinterpretation of Chinese architectural culture in the new architectural theory system, and learn the architectural concepts and methods of reinterpretation of Chinese architectural culture.

Course content
1. Overview and foundation of this course 2. Chinese architecture: western interpretation and the change of western architectural concept 3. Chinese architecture: Chinese interpretation and the establishment of Chinese architecture academic system 4. Wood structure system: the significance of Chinese construction culture 5. Residence and garden: the cultural significance of human settlement in China 6. Harmony of cosmology: the significance of Chinese urban culture 7. Discussion

---

研究生一年级
建筑理论研究 · 王骏阳
课程类型：必修
学时/学分：18学时/1学分

Graduate Program 1st Year
STUDIES OF ARCHITECTURAL THEORY · WANG Junyang
Type: Required Course
Study Period and Credits: 18 hours / 1 credit

课程介绍
本课程是西方建筑史研究生教学的一部分。主要涉及当代西方建筑界具有代表性的思想和理论，其主题包括历史主义、先锋建筑、批判理论、建构文化以及对当代城市的解读等。本课程大量运用图片资料，广泛涉及哲学、历史、艺术等领域，力求在西方文化发展的背景中呈现建筑思想和理论的相对独立性及关联性，理解建筑作为一种人类活动所具有的社会和文化意义，启发学生的理论思维和批判精神。

课程内容
第一讲：建筑理论概论
第二讲：数字化建筑与传统建筑学的分离与融合
第三讲：语言、图解、空间内容
第四讲："拼贴城市"与城市的观念
第五讲：建构与营造
第六讲：手法主义与当代建筑
第七讲：从主线历史走向多元历史之后的思考
第八讲：讨论

Course description
This course is a part of western architectural history teaching for graduate students. It mainly deals with the representative thoughts and theories in western architectural circles, including historicism, vanguard building, critical theory, tectonic culture and interpretation of contemporary cities etc.. Using a lot of pictures involving extensive fields including philosophy, history, art, etc., this course attempts to show the relative independence and relevance of architectural thoughts and theories under the development background of western culture, understand the social and cultural significance owned by architecture as human activities, and inspire students' theoretical thinking and critical spirit.

Course content
Lecture 1: Overview of architectural theories
Lecture 2: Separation and integration between digital architecture and traditional architecture
Lecture 3: Language, diagram and spatial content
Lecture 4: "Collage city" and concept of city
Lecture 5: Tectonics and Yingzao (Ying-Tsao)
Lecture 6: Mannerism and modern architecture
Lecture 7: Thinking after main-line history to diverse history
Lecture 8: Discussion

研究生一年级
建筑史研究 · 胡恒
课程类型：选修
学时/学分：18学时/1学分

Graduate Program 1st Year
STUDIES OF THE HISTORY OF ARCHITECTURE · HU Heng
Type: Elective Course
Study Period and Credits: 18 hours / 1 credit

教学目标
促进学生对历史研究的主题、方法、路径有初步的认识，通过具体的案例讲解使学生能够理解当代中国建筑史研究的诸多可能性。

课程内容
1. 图像与建筑史研究（1—文学、装置、设计）
2. 图像与建筑史研究（2—文学、装置、设计）
3. 图像与建筑史研究（3—绘画与园林）
4. 图像与建筑史研究（4—绘画、建筑、历史）
5. 图像与建筑史研究（5—文学与空间转译）
6. 方法讨论1
7. 方法讨论2

Teaching objectives
To promote students' preliminary understanding of the topic, method and approach of historical research. To make students understand the possibilities of contemporary study on history of Chinese architecture through specific cases.

Course content
1. Image and architectural history study (1-literature, device, design)
2. Image and architectural history study (2-literature, device, design)
3. Image and architectural history study (3- painting and garden)
4. Image and architectural history study (4-painting, architecture and history)
5. Image and architectural history study (5-literature and spatial transform)
6. Method discussion 1
7. Method discussion 2

研究生一年级
中国木建构文化研究 · 赵辰
课程类型：选修
学时/学分：18学时/1学分

Graduate Program 1st Year
STUDIES IN CHINESE WOODEN TECTONIC CULTURE · ZHAO Chen
Type: Elective Course
Study Period and Credits: 18 hours / 1 credit

教学目标
以木为材料的建构文化是世界各文明中的基本成分，中国的木建构文化更是深厚而丰富。在全球可持续发展要求之下，木建构文化必须得到重新的认识和评价。对于中国建筑文化来说，更具有再认识和再发展文化传统的意义。

课程内容
阶段一：理论基础——对全球木建构文化的重新认识
阶段二：中国木建构文化的原则和方法（讲座与工作室）
阶段三：中国木建构的基本形——从家具到建筑（讲座与工作室）
阶段四：结构造型的发展和木建构的现代化（讲座）
阶段五：建造实验的鼓动（讲座与工作室）

Teaching objectives
The wood – based construction culture is the basic component of all civilizations in the world, and Chinese wood construction culture is profound and abundant. Under the requirement of global sustainable development, wood construction culture must be re-recognized and evaluated. For Chinese architectural culture, it is of great significance to re-recognize and re-develop the cultural tradition.

Course content
Stage 1: theoretical basis—re-understanding of global wood construction culture
Stage 2: principles and methods of Chinese wood construction culture (lectures and studios)
Stage 3: the basic shape of Chinese wood construction—from furniture to architecture (lectures and studios)
Stage 4: development of structural modeling and modernization of wood construction (lectures)
Stage 5: agitation of construction experiment (lectures and studios)

建筑技术课程
ARCHITECTURAL TECHNOLOGY COURSES

本科二年级
CAAD 理论与实践・吉国华 傅筱 万军杰
课程类型：选修
学时 / 学分：36 学时 / 2 学分

Undergraduate Program 2nd Year
THEORY AND PRACTICE OF CAAD
• JI Guohua, FU Xiao, WAN Junjie
Type: Elective Course
Study Period and Credits: 36 hours / 2 credits

课程介绍
  在现阶段的 CAD 教学中，强调了建筑设计在建筑学教学中的主干地位，将计算机技术定位于绘图工具，本课程就是帮助学生可以尽快并且熟练地掌握如何利用计算机工具进行建筑设计的表达。课程中整合了 CAD 知识、建筑制图知识以及建筑表现知识，将传统 CAD 教学中教会学生用计算机绘图的模式向教会学生用计算机绘制有形式感的建筑图的模式转变，强调准确性和表现力作为评价 CAD 学习的两个最重要指标。
  本课程的具体学习内容包括：
  1. 初步掌握 AutoCAD 软件和 SketchUP 软件的使用，能够熟练完成二维制图和三维建模的操作；
  2. 掌握建筑制图的相关知识，包括建筑投影的基本概念、平立剖面、轴测、透视和阴影的制图方法和技巧；
  3. 图面效果表达的技巧，包括黑白线条图和彩色图纸的表达方法和排版方法。

Course description
The core position of architectural design is emphasized in the CAD course. The computer technology is defined as drawing instrument. The course helps students learn how to make architectural presentation using computers fast and expertly. The knowledge of CAD, architectural drawing and architectural presentation are integrated into the course. The traditional mode of teaching students to draw in CAD course will be transformed into teaching students to draw architectural drawing with sense of forms. The precision and expression will be emphasized as two most important factors to estimate the teaching effect of CAD course.
Contents of the course include:
1. Use AutoCAD and SketchUP to achieve the 2D drawing and 3D modeling expertly.
2. Learn relational knowledge of architectural drawing, including basic concepts of architectural projection, drawing methods and skills of plan, elevation, section, axonometry, perspective and shadow.
3. Skills of presentation, including the methods of expression and lay out using mono and colorful drawings

---

本科三年级
建筑技术（一）・傅筱 李清朋
课程类型：必修
学时 / 学分：36 学时 / 2 学分

Undergraduate Program 3rd Year
ARCHITECTURAL TECHNOLOGY 1 • FU Xiao, LI Qingpeng
Type: Required Course
Study Period and Credits: 36 hours / 2 credits

课程介绍
  本课程是建筑学专业本科生的专业主干课程。本课程的任务主要是以建筑师的工作性质为基础，讨论一个建筑生成过程中最基本的三大技术支撑（结构、构造、施工）的原理性知识要点，以及它们在建筑实践中的相互关系。

Course description
The course is a major course for the undergraduate students of architecture. The main purpose of this course is based on the nature of the architect's work, to discuss the principle knowledge points of the basic three technical supports in the process of generating construction (structure, construction, execution), and their mutual relations in the architectural practice.

本科三年级
建筑技术（二）声光热・吴蔚
课程类型：必修
学时 / 学分：36 学时 / 2 学分

Undergraduate Program 3rd Year
ARCHITECTURAL TECHNOLOGY 2 SOUND, LIGHT AND HEAT • WU Wei
Type: Required Course
Study Period and Credits: 36 hours / 2 credits

课程介绍
  本课程是针对三年级学生所设计，课程介绍了建筑热工学、建筑光学、建筑声学中的基本概念和基本原理，使学生能掌握建筑的热环境、声环境、光环境的基本评估方法，以及相关的国家标准。学生完成学业后在此方向上能阅读相关书籍，具备在数字技术方法等相关资料的帮助下，完成一定的建筑节能设计的能力。

Course description
Designed for the Grade 3rd students, this course introduces the basic concepts and basic principles in architectural thermal engineering, architectural optics and architectural acoustics, so that the students can master the basic methods for the assessment of building's thermal environment, sound environment and light environment as well as the related national standards. After graduation, the students will be able to read the related books regarding these aspects, and have the ability to complete certain building energy efficiency design with the help of the related digital techniques and methods.

本科三年级
建筑技术（三）水电暖・吴蔚
课程类型：必修
学时 / 学分：36 学时 / 2 学分

Undergraduate Program 3rd Year
ARCHITECTURAL TECHNOLOGY 3 WATER, ELECTRICITY AND HEATING • WU Wei
Type: Required Course
Study Period and Credits: 36 hours / 2 credits

课程介绍
  本课程是针对南京大学建筑与城市规划学院本科三年级学生所设计。课程介绍了建筑给水排水系统、采暖通风与空气调节系统、电气工程的基本理论、基本知识和基本技能，使学生能熟练地阅读水电、暖通工程图，熟悉水及和消防的设计、施工规范，了解燃气供应、安全用电及建筑防火、防雷的初步知识。

Course description
This course is an undergraduate class offered in the School of Architecture and Urban Planning, Nanjing University. It introduces the basic principle of the building service systems, the technique of integration amongst the building services and the building. Throughout the course, the fundamental importance to energy, ventilation, air-conditioning and comfort in buildings are highlighted.

研究生一年级
传热学与计算流体力学基础・郜志
课程类型：选修
学时 / 学分：36 学时 / 2 学分

Graduate Program 1st Year
FUNDAMENTALS OF HEAT TRANSFER AND COMPUTATIONAL FLUID DYNAMICS • GAO Zhi
Type: Elective Course
Study Period and Credits: 36 hours / 2 credits

课程介绍
  本课程的主要任务是使建筑学 / 建筑技术学专业的学生掌握传热学和计算流体力学的基本概念和基本知识，通过课程教学，使学生熟悉传热学中导热、对流和辐射的经典理论，并了解传热学和计算流体力学的实际应用和最新研究进展，为建筑能源和环境系统的计算和模拟打下坚实的理论基础。教学中尽量简化传热学和计算流体力学经典课程中复杂公式的推导过程，而着重于如何解决建筑能源与建筑环境中涉及流体流动和传热的实际应用问题。

Course description
This course introduces students majoring in Building Science and Engineering / Building Technology to the fundamentals of heat transfer and computational fluid dynamics (CFD). Students will study classical theories of conduction, convection and radiation when heat transfers, and learn advanced research developments of heat transfer and CFD. The complex mathematics and physics equations are not emphasized. It is desirable that for real-case scenarios students will have the ability to analyze flow and heat transfer phenomena in building energy and environment systems.

研究生一年级
GIS 基础与应用・童滋雨
课程类型：必修
学时 / 学分：18 学时 / 1 学分

Graduate Program 1st Year
CONCEPT AND APPLICATION OF GIS • TONG Ziyu
Type: Required Course
Study Period and Credits: 18 hours / 1 credit

课程介绍
  本课程的主要目的是让学生理解 GIS 的相关概念以及 GIS 对城市研究的意义，并能够利用 GIS 软件对城市进行分析和研究。

Course description
This course aims to enable students to understand the related concept of GIS and the significance of GIS to urban research, and to be able to use GIS software to carry out urban analysis and research.

研究生一年级
建筑环境学与设计 · 邰志
课程类型：必修
学时 / 学分：36 学时 /2 学分

Graduate Program 1st Year
ARCHITECTURAL ENVIROMENTAL SCIENCE AND DESIGN
• GAO Zhi
Type: Required Course
Study Period and Credits:36 hours / 2 credits

课程介绍
　　本课程的主要任务是使建筑学 / 建筑技术学专业的学生掌握建筑环境的基本概念，学习建筑与城市热湿环境、风环境和空气质量的基础理论。通过课程教学，使学生熟悉城市微气候等理论，并了解人体对热湿环境的反应，掌握建筑环境学的实际应用和最新研究进展，为建筑能源和环境系统的测量与模拟打下坚实的基础。

Course description
This course introduces students majoring in Building Science and Engineering / Building Technology to the fundamentals of built environment. Students will study classical theories of built / urban thermal and humid environment, wind environment and air quality. Students will also familiarize urban micro environment and human reactions to thermal and humid environment. It is desirable that students will have the ability to measure and simulate building energy and environment systems based upon the knowledge of the latest development of the study of built environment.

研究生一年级
材料与建造 · 冯金龙
课程类型：选修
学时 / 学分：18 学时 /1 学分

Graduate Program 1st Year
MATERIAL AND CONSTRUCTION
• FENG Jinlong
Type: Elective Course
Study Period and Credits:18 hours / 1 credit

课程介绍
　　本课程将介绍现代建筑技术的发展过程，论述现代建筑技术及其美学观念对建筑设计的重要作用；探讨由材料、结构和构造方式所形成的建筑建造的逻辑方式；研究建筑形式产生的物质技术基础，诠释现代建筑的建构理论与研究方法。

Course description
It introduces the development process of modern architecture technology and discusses the important role played by the modern architecture technology and its aesthetic concept in the architectural design. It explores the logical methods of construction of the architecture formed by materials, structure and construction. It studies the material and technical basis for the creation of architectural form, and interprets the construction theory and research method for modern architecture.

研究生一年级
计算机辅助技术 · 吉国华
课程类型：选修
学时 / 学分：36 学时 /2 学分

Graduate Program 1st Year
COMPUTER AIDED DESIGN • JI Guohua
Type: Elective Course
Study Period and Credits:36 hours / 2 credits

课程介绍
　　随着计算机辅助建筑设计技术的快速发展，当前数字技术在建筑设计中的角色逐渐从辅助绘图转向了真正的辅助设计，并引发了设计的革命和建筑的形式创新。本课程讲授Grasshopper 参数化编程建模方法以及相关的几何知识，让学生在掌握参数化编程建模技术的同时，增强以理性的过程思维方式分析和解决设计问题的能力，为数字建筑设计和数字建造打下必要的基础。
　　基于 Rhinoceros 的算法编程平台 Grasshopper 的参数化建模方法，讲授内容包括各类运算器的功能与使用、图形的生成与分析、数据的结构与组织、各类建模的思路与方法，以及相应的数学与计算机编程知识。

Course description
The course introduces methods of Grasshopper parametric programming and modeling and relevant geometric knowledge. The course allows students to master these methods, and enhance ability to analyze and solve designing problems with rational thinking at the same time, building necessary foundation for digital architecture design and digital construction.
In this course, the teacher will teach parametric modeling methods based on Grasshopper, a algorithmic programming platform for Rhinoceros, including functions and application of all kinds of arithmetic units, pattern formation and analysis, structure and organization of data, various thoughts and methods of modeling, and related knowledge of mathematics and computer programming.

研究生一年级
建筑体系整合 · 吴蔚
课程类型：必修
学时 / 学分：18—36 学时 /1—2 学分

Graduate Program 1st Year
BUILDING SYSYTEM INTEGRATION • WU Wei
Type: Required Course
Study Period and Credits: 18-36 hours / 1-2 credits

课程介绍
　　本课程是从建筑各个体系整合的角度来解析建筑设计。首先，课程介绍了建筑体系整合的基本概念、原理及其美学观念；然后具体解读以上各个设计元素在整个建筑体系中所扮演的角色及其影响力，了解建筑各个系统之间的互相联系和作用；最后，以全球的环境问题和人类生存与发展为着眼点，引导同学们重新审视和评判我们奉为信条的设计理念和价值系统。本课程着重强调建筑设计需要了解不同学科和领域的知识，熟悉各工种之间的配合和协调。

Course description
A building is an assemblage of materials and components to obtain a shelter from external environment with a certain amount of safety so as to provide a suitable internal environment for physiological and psychological comfort in an economical manner. This course examines the role of building technology in architectural design, shows how environmental concerns have shaped the nature of buildings, and takes a holistic view to understand the integration of different building systems. It employs total building performance which is a systematic approach, to evaluate the performance of various sub-systems and to appraise the degree of integration of the sub-systems.

研究生一年级
数字建筑设计 · 吉国华　童滋雨
课程类型：选修
学时 / 学分：36 学时 /2 学分

Graduate Program 1st Year
DIGITAL ARCHITECTURAL DESIGN
• JI Guohua, TONG Ziyu
Type: Elective Course
Study Period and Credits: 36 hours / 2 credits

课程介绍
　　编程技术是数字建筑的基础，本课程主要讲授Grasshopper 脚本编程和 Processing 编程，让学生在掌握代码编程基础技术的同时，增强以理性的过程思维方式分析和解决设计问题的能力，逐步掌握数字设计的方法，为数字设计和建造课程打好基础。

Course description
Programming technology is the foundation of digital architecture, this course mainly teaches Grasshopper script programming and Processing programming, so that students can master the basic technology of code programming, at the same time, enhance the ability to analyze and solve design problems with rational process thinking, gradually master the method of digital design, and lay a good foundation for the course of digital design and construction.

研究生一年级
建设工程项目管理 · 谢明瑞
课程类型: 选修
学时 / 学分: 36 学时 /2 学分

Undergraduate Program 1st Year
MANAGEMENT OF CONSTRUCTION PROJECT ·XIE Mingrui
Type: Elective Course
Study Period and Credits: 36 hours / 2 credits

课程介绍
　　帮助学生系统掌握建设工程项目管理的基本概念、理论体系和管理方法, 了解建筑规划设计在建设工程项目中的地位、特点和重要性。
　　延展建筑学专业学生基本知识结构层面, 拓展学生的发展方向。

Course description
To help students systematically master the basic concept, theoretical system and management method of construction engineering project management, understand the position, characteristics and importance of architectural planning design in the construction engineering project.
To extend the basic knowledge structure level of students majoring in architecture, extend the development direction of students.

研究生一年级
建筑环境学与设计 · 尤伟 郜志
课程类型: 必修
学时 / 学分: 18 学时 /1 学分

Undergraduate Program 1st Year
ARCHITECTURAL ENVIRONMENTAL SCIENCE AND DESIGN · YOU Wei, GAO Zhi
Type: Required Course
Study Period and Credits:18 hours / 1 credit

课程介绍
　　本课程是基于建筑环境学课程的设计实践课程, 意在将建筑环境学课程的理论知识通过设计案例的练习加以运用, 加深对建筑环境学知识的理解, 并训练如何通过设计优化营造良好室内环境品质。
　　课程分为授课和案例设计练习两部分, 授课部分介绍目前关于被动式设计研究成果、工程实践案例中的被动式设计方法以及软件模拟分析技术; 案例设计练习教授学生学习基于性能评估的优化设计方法, 选取学生较为熟悉的住宅、幼儿园等体量较小的建筑类型作为设计优化对象, 通过软件分析发现现有的室内环境设计不足, 并基于现有研究成果提出优化策略, 最后通过软件模拟加以验证。课程要求学生将建筑环境学课程所学知识用于本设计课程的室内环境品质的量化及控制。本课程着重训练建筑设计与环境工程学科知识的配合。

Course description
This course is a design practice course based on the course of building environment, aiming to apply the theoretical knowledge of the building environment course through the practice of design cases, so as to deepen the understanding of the knowledge of building environment, and train how to create a good indoor environment quality through design optimization.
The course is divided into two parts: teaching and case design practice. The teaching part introduces current research results of passive design, passive design methods in engineering practice cases and software simulation analysis technology.Case design practice teaches students to study optimal design method based on performance evaluation, choose the residence, kindergarten and other buildings with small volume that students are more familiar with as the design optimization objects, find existing deficiency in indoor environment design through software analysis, propose optimization strategies according to existing research result, and finally verify through software simulation. The course requires students to apply what they have learned in the building environment course to the quantification and control of indoor environment quality in this design course.This course focuses on the integration of architectural design and environmental engineering knowledge.

研究生一年级
建筑学中的技术人文主义 · 窦平平
课程类型: 必修
学时 / 学分: 36 学时 /2 学分

Undergraduate Program 1st Year
TECHNOLOGY OF HUMANISM IN ARCHITECTURE · DOU Pingping
Type: Required Course
Study Period and Credits:36 hours / 2 credits

课程介绍
　　课程详尽阐释了为满足建筑的多方需求而投入的技术探索和人文关怀。课程包括四大版块, 共 16 个主题讲座, 以案例精述的形式引介相关建筑师和学者的作品和理论。希望培养学生对建筑学中的技术议题进行批判性和人文主义的深入理解。

Course description
This course elaborates the technological endeavors and humanistic concern in fulfilling the multifaceted architectural demands. It takes shape in a series of sixteen theme lectures, grouped in four sections, introducing prominent architects and scholars through richly illustrated case studies and interpretations. It aims to nurture the students with a critical and humanistic understanding of the role of technology playing in the discipline of architecture.

其他
MISCELLANEA

讲座
Lectures

硕士学位论文列表
List of Thesis for Master Degree

研究生姓名	研究生论文标题	导师姓名
刘宛莹	基于结构设备一体化的高层办公建筑异形梁模板设计研究	丁沃沃
施少鎏	基于结构设备一体化的高层办公建筑异形梁设计研究	丁沃沃
范 勇	基于GIS平台的城市路网立交特征图示化研究	丁沃沃
林瑜洋	运动视角下街道空间特征描述性指标研究	丁沃沃
倪 铮	街廓地块指标与街廓立面轮廓的关联性研究	丁沃沃
吴慧敏	基于工业化竹材应用的书架与建筑的集成结构设计研究	赵 辰
张文轩	晋北堡城的堡墙建构与再生设计研究——以杀虎堡为例	赵 辰
潮书镛	城水互动——明代以来南京城东南水系演变研究	赵 辰
陈健楠	近代中国建筑若干外来概念跨文化转译研究（1910s—1940s）	赵 辰
方园园	浅析杨廷宝"得体合宜"建筑观——以其钢筋混凝土结构大屋顶作品为例	赵 辰
刘颖琦	立足于青年群体的南京公共租赁住房设计策略研究	王骏阳
王云珂	中国近代工业建筑研究发展史初探	王骏阳
冯时雨	基于触控设备的建筑草图工具开发初探	吉国华
金沛沛	风格迁移和条件对抗生成网络在城市肌理生成中的应用研究	吉国华
梁晓蕊	基于消费者行为的多智能模拟在商业规划中的应用	吉国华
李 让	曲面的无扭转节点网格重构	吉国华
李 雅	基于正交剖分Voronoi图的建筑平面生成方法	吉国华
罗文馨	基于三角形与多边形混合的曲面平板化重构研究	吉国华
黎飞鸣	编织表皮的设计研究——以余杭东湖书院项目为例	张 雷
俞言曦（林晨晨）	基于城市文脉的公共空间设计研究——以石家庄中央商务区展示中心为例	张 雷
唐 萌	在地性乡村公共建筑营造策略研究——以李庄安石文化中心为例	张 雷
王慧文	基于乡村公共空间营造的乡村社区中心设计——以桐庐县旧县村文化礼堂为例	张 雷
王 顺	体验式商业建筑空间研究——以南京孙家祠堂商业项目为例	张 雷
杨云睿	石家庄中央商务区展示中心设计及UHPC建筑设计研究	张 雷
蒋 健	南京市永乐南路政务服务中心设计——政务服务类建筑的功能组织和空间设计研究	冯金龙
李 天	苏州深时数字地球国际卓越研究中心设计——科创办公建筑交互空间设计研究	冯金龙
秦 岭	南京大学苏州校区公共科研楼设计——学科群建筑模块化设计策略研究	冯金龙
张路薇	南京永乐南路社区中心设计——社区中心连续空间营造策略研究	冯金龙
时 远	建筑师视角下建筑设计全流程管理要素研究——使用需求转化为设计成果的控制策略探索	冯金龙、谢明瑞
王少君	消费转型背景下的主题商业设计和场景营造——以扬州图书馆东侧地块项目为例	冯金龙、谢明瑞
陈鹏远	基于新乡土主义的乡村公共建筑设计策略研究——以盐城穆沟村公共组团设计为例	周 凌
林 宇	钢木组合结构轻型建筑的结构骨架和构造研究——以南京金牛湖原舍公区为例	周 凌
孙晓雨	驿站建筑设计研究——以南京市溧水区回峰山驿站设计为例	周 凌
孙媛媛	基于室外风环境模拟的环渤海地区多层住宅布局策略研究	周 凌
尹子晗	基于室外风环境模拟的长三角地区多层住宅布局策略研究	周 凌
张珊珊	从格伦·马库特住宅建造体系到轻型建造体系设计应用研究	周 凌

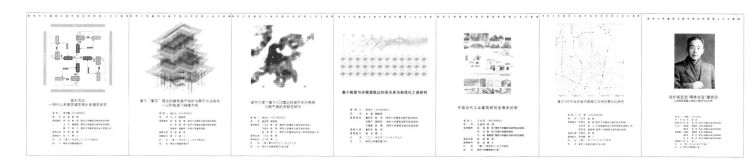

研究生姓名	研究生论文标题	导师姓名
陈妍霓	基础的设计与表达——基于《建构设计——材料·构造·结构》教材的研究	傅筱
胡名卉	预制混凝土外墙挂板、洞口构造设计与表达——基于《建构设计——材料·构造·结构》教材的研究	傅筱
柳妍	干挂雨幕系统的构造与表达——基于《建构设计——材料·构造·结构》教材的研究	傅筱
张涛	建筑垂直交通体系的人性化设计——基于《建构设计——材料·构造·结构》教材的研究	傅筱
张文欣	现代木构建筑的外围护结构的构造与表达——基于《建构设计——材料·构造·结构》的专题研究	傅筱
张雅翔	女儿墙的构造与表达——基于《建构设计——材料·构造·结构》教材的研究	傅筱
陈晓	用于纪念性博览建筑空间经验评估的身体影像分析方法研究	鲁安东
杜孟泽杉	空间中的身体——电影建筑学视角下的巴斯特·基顿影像空间研究	鲁安东
聂书琪	医学身体视角下现代主义时期为医生而设计的住宅研究	鲁安东
徐琳茜	面向城市品牌塑造的场所网络运作机制和设计策略研究	鲁安东
张思琪	"文学地图"绘制方法研究——以《魏特琳日记》为例	鲁安东
李斓珺	明代南京西园研究——基于文本、图像的解读与再现	胡恒
王熙昀	巴尔达萨雷佩鲁齐研究——论文艺复兴建筑师的罗马意识	胡恒
黄瑞安	作为社区触媒的基础设施赋能——宁芜铁路秦虹段改造更新设计	华晓宁
李星儿	乡村农产品观光工厂的设计研究——常州丁家村特色食品观光工厂方案设计	华晓宁
刘洋	石臼湖—固城湖圩区团块状圩村更新研究——南京市高淳区钱家村改造更新设计	华晓宁
刘伟	改革开放以来苏中地区农村自建住宅形态研究	华晓宁
陈启宁	基于视觉与非视觉效应的采光多目标优化工具研究	吴蔚
郭鑫	典型布局住区的交通污染扩散特性及其对建筑通风策略和能耗的影响	郜志
徐志超	行驶车辆对不同形态街道峡谷内空气流动的影响	郜志
张悦	基于城市风环境模拟的集成于建筑设计平台CFD工具测评研究	郜志
夏心雨	基于多源数据与机器学习的城市空间布局生成方法研究	童滋雨
严华东	挤出多面体变形特征和应用研究	童滋雨
郑航	基于BIM模型的IndoorGML模型生成与应用	童滋雨
周珏伦	城市尺度下基于LCZ理论的城市形态格局与微气候的关联性研究	童滋雨
李家祥	气候缓冲层在公共建筑设计中的应用——以DDE国际会议交流中心优化设计为例	胡友培
李晓楠	瑞金市枣米巷历史街区渐进式保护更新设计研究	胡友培
迟铄雯	地块规模、地块指标、建筑肌理的关联性研究	胡友培
刘恺丽	当代城市设计中景观生态架构的议题与方法	胡友培
刘婧雯	"生产·生活·生态"一体的创意农业建筑设计研究	窦平平
杨璐	行为与情感数据的空间化表达——一种多学科链接的可视化路径	窦平平
黄健佳	空间生产理论视角下"浙江庆和昌记支店旧址"的多重价值研究	冷天
李谷羽	基于历史信息转译的斗鸡闸地块空间信息要素演变研究	冷天
陈红云	基于"重写"理论的建筑遗产保护与展示方法研究——以钓鱼城飞舄楼为例	冷天
刘晓芬	南京大学北大楼塔楼建设始末初探	冷天
王春燕	黔东北凯望村住宅中堂屋空间的解析及其再现研究	冷天

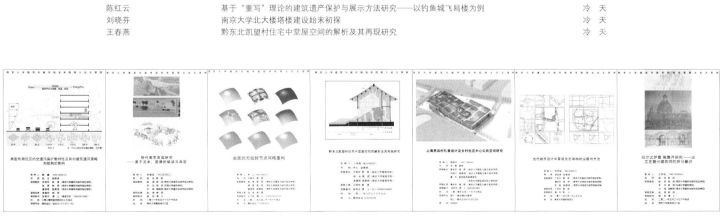

在校学生名单
List of Students

本科生 Undergraduate

2017级学生 / Students 2017

卞直瑞 BIAN Zhirui	樊力立 FAN Lili	刘畅 LIU Chang	沈葛梦欣 SHEN Gemengxin	张凯莉 ZHANG Kaili
卜子睿 BU Zirui	甘静雯 GAN Jingwen	龙沄 LONG Yun	沈晓燕 SHEN Xiaoyan	周金雨 ZHOU Jinyu
陈佳晨 CHEN Jiachen	顾天奕 GU Tianyi	陆柚余 LU Youyu	孙萌 SUN Meng	周慕尧 ZHOU Muyao
陈露茜 CHEN Luxi	韩小如 HAN Xiaoru	马路遥 MA Luyao	孙瀚 SUN Han	朱菁菁 ZHU Jingjing
陈雨涵 CHEN Yuhan	焦梦雅 JIAO Mengya	马子昂 MA Ziang	杨帆 YANG Fan	朱雅芝 ZHU Yazhi
程科懿 CHENG Keyi	李心彤 LI Xintong	彭洋 PENG Yang	杨佳锟 YANG Jiakun	达热亚·阿吾斯哈力 Dareya Awusihali
董一凡 DONG Yifan	林易谕 LIN Yiyu	尚紫鹏 SHANG Zipeng	杨乙彬 YANG Yibin	

2018级学生 / Students 2018

包诗贤 BAO Shixian	林济武 LIN Jiwu	邱雨欣 QIU Yuxin	肖郁伟 XIAO Yuwei	张同 ZHANG Tong
陈锐娇 CHEN Ruijiao	刘瑞翔 LIU Ruixiang	沈新洁 SHEN Jie	熊浩宇 XIONG Haoyu	张新雨 ZHANG Xinyu
冯德庆 FENG Deqing	刘湘菲 LIU Xiangfei	宋佳艺 SONG Jiayi	徐颖 XU Ying	周宇阳 ZHOU Yuyang
顾嵘健 GU Rongjian	陆麒竹 LU Qizhu	孙穆群 SUN Muqun	薛云龙 XUE Yunlong	阿尔申·巴特尔江 Aershen Bateerjiang
顾祥姝 GU Xiangshu	罗宇豪 LUO Yuhao	田靖 TIAN Jing	杨朵 YANG Duo	
何旭 HE Xu	倪梦琪 NI Mengqi	田舒琳 TIAN Shulin	喻姝凡 YU Shufan	
李逸凡 LI Yifan	牛乐乐 NIU Lele	吴高鑫 WU Gaoxin	张百慧 ZHANG Baihui	

2019级学生 / Students 2019

高禾雨 GAO Heyu	石珂千 SHI Keqian	周昌赫 ZHOU Changhe
高赵龙 GAO Zhaolong	唐诗诗 TANG Shishi	上原舜平 SHANGYUAN Shunping
顾靓 GU Liang	王思戎 Wang Sirong	麦吾兰江·穆合塔尔 Maiwulanjiang Muhetar
黄辰逸 HUANG Chenyi	王智坚 WANG Zhijian	
黄小东 HUANG Xiaodong	王梓蔚 WANG Ziwei	
黄煜东 HUANG Yudong	袁泽 YUAN Ze	
邱雨婷 QIU Yuting	张楚杭 ZHANG Chuhang	

2020级学生 / Students 2020

李若松 Li Ruosung	陈浏毓 Chen Liuyu	沈至文 Shen Zhiwen	陈玎 Chen Cheng
钱梦南 Qian Mengnan	陆星宇 Lu Xingyu	李静怡 Li Jingyi	张伊儿 Zhang Yi'er
袁欣鹏 Yuan Xinpeng	顾林 Gu Lin	李沛熹 Li Peixi	何德林 He Delin
刘晓斌 Liu Xiaobin	吴嘉文 Wu Jiawen	杨曦睿 Yang Xirui	王天歌 Wang Tian'ge
华羽纶 Hua Yuguan	张嘉木 Zhang Jiamu	高晴 Gao Qing	陈璇霖 Chen Xuanlin
刘珩歆 Liu Hengxin	黄淑睿 Huang Shurui	刘卓然 Liu Zhuoran	
陈沈婷 Chen Shenting	孙昊天 Sun Haotian	王天赐 Wang Tianci	

研究生 Postgraduate

曹焱 CAO Yan	方园园 FANG Yuanyuan	刘霄 LIU Xiao	王云珂 WANG Yunke	陈鹏远 CHEN Pengyuan	蒋健 JIANG Jian	梁晓蕊 LIANG Xiaorui	邵夏梦 SHAO Xiameng	王顺 WANG Shun	张思琪 ZHANG Siqi
潮书镛 CHAO Shuyong	郭鑫 GUO Xin	刘晓芬 LIU Xiaofen	徐琳茜 XU Linxi	陈霄 CHEN Xiao	金沛沛 JIN Peipei	林晨晨 LIN Chenchen	施少鋆 SHI Shaojun	王维依 WANG Weiyi	张涛 ZHANG Tao
陈红云 CHEN Hongyun	李斓珺 LI Lanjun	刘颖琦 LIU Yinqi	徐志超 XU Zhichao	陈妍霓 CHEN Yanni	黎飞鸣 LI Feiming	林宇 LIN Yu	时远 SHI Yuan	吴慧敏 WU Huimin	张文欣 ZHANG Wenxin
陈健楠 CHEN Jiannan	李让 LI Rang	罗文馨 LUO Wenxin	严华东 YAN Huadong	杜孟泽杉 DUMENG Zeshan	李谷羽 LI Guyu	刘靖雯 LIU Jingwen	孙晓雨 SUN Xiaoyu	夏心雨 XIA Xinyu	张文轩 ZHANG Wenxuan
陈启宁 CHEN Qining	李雅 LI Ya	倪铮 NI Zheng	杨璐 YANG Lu	冯时雨 FENG Shiyu	李家祥 LI Jiaxiang	刘宛莹 LIU Wanying	孙媛媛 SUN Yuanyuan	杨云睿 YANG Yunrui	张雅翔 ZHANG Yaxiang
陈紫葳 CHEN Ziwei	林瑜洋 LIN Yuyang	聂书琪 NIE Shuqi	张悦 ZHANG Yue	胡名卉 HU Minghui	李天 LI Tian	刘洋 LIU Yang	唐萌 TANG Meng	尹子晗 YIN Zihan	张钰 ZHANG Yu
迟铄雯 CHI Shuowen	刘恺丽 LIU Kaili	王春燕 WANG Chunyan	郑航 ZHENG Hang	黄健佳 HUANG Jianjia	李晓楠 LI Xiaonan	柳妍 LIU Yan	王慧文 WANG Huiwen	张路薇 ZHANG Luwei	左斌 ZUO Bin
范勇 FAN Yong	刘伟 LIU Wei	王熙昀 WANG Xiyun	周珏伦 ZHOU Juelun	黄瑞安 HUANG Ruian	李星儿 LI Xinger	秦岭 QIN Ling	王少君 WANG Shaojun	张珊珊 ZHANG Shanshan	周郅 ZHOU Zhi

卞真 BIAN Zhen	皇甫子玥 HUANGFU Ziyue	吕文倩 LV Wenqian	魏雪仪 WEI Xueyi	岑国桢 CEN Guozhen	冯杨帆 FENG Yangfan	黄晓寻 HUANG Xiaoxun	刘贺 LIU He	王梦兰 WANG Menglan	郑经纬 ZHENG Jingwei
陈莉莉 CHEN Lili	孔严 KONG Yan	明文静 MING Wenjing	辛宇 XIN Yu	车娟娟 CHE Juanjuan	傅婷婷 FU Tingting	匡鑫 KUANG Xin	罗逍遥 LUO Xiaoyao	王若辰 Wang Ruochen	王鹏程 WANG Pengcheng
陈星雨 CHEN Xingyu	李昂 LI Ang	濮文睿 PU Wenrui	杨子媛 YANG Ziyuan	陈志凡 CHEN Zhifan	戈可辰 GE Kechen	况赫 KUANG He	莫默 MO Mo	王赛施 WANG Saishi	朱维韬 ZHU Weitao
陈一帆 CHEN Yifan	李博雅 LI Boya	沈育辉 SHEN Yuhui	袁琴 YUAN Qin	程绪 CHENG Xu	谷雨阳 GU Yuyang	李乐 LI Yue	史鑫尧 SHI Xinyao	王子涵 WANG Zihan	朱晓晨 ZHU Xiaochen
陈玉珊 CHEN Yushan	李昌曦 LI Changxi	孙其 SUN Qi	张含 ZHANG Han	戴添趣 DAI Tianqu	顾梦婕 GU Mengjie	李欣仪 LI Xinyi	宋晓宇 SONG Xiaoyu	温琳 WEN Lin	
程薏 CHENG Yi	李嘉伟 LI Jiawei	谭锦楠 TAN Jinnan	赵彤 ZHAO Tong	丁展图 DING Zhantu	何璇 HE Xuan	李雪 LI Xue	谭路路 TAN Lulu	翁昕 WENG Xin	
洪静 HONG Jing	李伊萌 LI Yimeng	王新强 WANG Xinqiang	周诗琪 ZHOU Shiqi	董青 DONG Qing	胡应航 HU Yinghang	李芸梦 LI Yunmeng	王家洲 WANG Jiazhou	臧哲 ZANG Zhe	
侯自忠 HOU Zizhong	梁颖 LIANG Yin	王旭 WANG Xu	朱硕 ZHU Shuo	范嫣琳 FANG Yanlin	黄菲 HUANG Fei	廖伟平 LIAO Weiping	王锴 WANG Kai	张尊 ZHANG Zun	

陈婧秋 CHEN Jingqiu	李晓云 LI Xiaoyun	王琪 WANG Qi	张云松 ZHANG Yunsong	程慧 CHENG Hui	赖泽贤 LAI Zexian	丘雨辰 QIU Yuchen	翁鸿祎 WENG Hongyi	张梦冉 ZHANG Mengran	
陈茜 CHEN Qian	林之茜 LIN Zhiqian	吴子豪 WU Zihao	赵琳芝 ZHAO Linzhi	仇佳豪 QIU Jiahao	李昂 LI Ang	任钰佳 REN Yujia	谢文俊 XIE Wenjun	张塑琪 ZHANG Suqi	
陈颖 CHEN Ying	刘雪寒 LIU Xuehan	谢菲 XIE Fei	周宇飞 ZHOU Yufei	崔晓伟 CUI Xiaowei	刘亲贤 LIU Qinxian	盛泽铭 SHENG Zeming	徐佳楠 XU Jianan	赵济宁 ZHAO Jining	
龚泰丹 GONG Tairan	刘奕孜 LIU Yizi	邢雨辰 XING Yuchen	朱激清 ZHU Jiqing	丁嘉欣 DING Jiaxin	刘雨田 LIU Yutian	宋贻泽 SONG Yize	杨淑钊 YANG Shuchuan	赵亚迪 ZHAO Yadi	
胡永裕 HU Yongyu	马致远 MA Zhiyuan	徐福锁 XU Fusuo	白珂嘉 BAI Kejia	丁明昊 DING Minghao	罗紫娟 LUO Zijuan	孙杰 SUN Jie	于瀚清 YU Hanqing	赵子文 ZHAO Ziwen	
黄翊婕 HUANG Yijie	庞馨怡 PANG Xinyi	许龄 XU Ling	曾敬琪 ZENG Jingqi	方奕璇 FANG Yixuan	吕广彤 LV Guangtong	唐敏 TANG Min	于文爽 YU Wenshuang	周理洁 ZHOU Lijie	
蒋哲 JIANG Zhe	邵桐 SHAO Tong	杨东来 YANG Donglai	陈铭行 CHEN Mingxing	费元丽 FEI Yuanli	马丹艺 MA Danyi	王明珠 WANG Mingzhu	于智超 YU Zhichao	朱辰浩 ZHU Chenhao	
雷畅 LEI Chang	王路 WANG Lei	杨岚 YANG Lan	陈予婧 CHEN Yujing	冯智 FENG Zhi	潘晴 PAN Qing	王瑞蓬 WANG ruipeng	余沁葼 YU qinman	朱凌云 ZHU Lingyun	
李倩 LI Qian	王路 WANG Lu	杨瑞侃 YANG ruikan	陈宇帆 CHEN Yufan	龚豪辉 GONG Haohui	秦钢强 QIN Gangqiang	王雨嘉 WANG Yujia	袁振香 YUAN Zhenxiang	朱孟阳 ZHU Mengyang	

图书在版编目（CIP）数据

南京大学建筑与城市规划学院建筑系教学年鉴. 2020—2021 / 胡友培编. -- 南京：东南大学出版社，2022.6
ISBN 978-7-5766-0142-8

Ⅰ.①南… Ⅱ.①胡… Ⅲ.①南京大学-建筑学-教学研究-2020-2021-年鉴 Ⅳ.①TU-42

中国版本图书馆CIP数据核字（2022）第103973号

编 委 会：丁沃沃　赵　辰　吉国华　周　凌　胡友培
版面制作：胡友培　王峥涛　燕海南　曹舒琪　王　琪　朱激清
责任编辑：姜　来　魏晓平
责任校对：张万莹
封面制作：毕　真
责任印制：周荣虎

南京大学建筑与城市规划学院建筑系教学年鉴 2020—2021

Nanjing Daxue Jianzhu Yu Chengshi Guihua Xueyuan Jianzhuxi Jiaoxue Nianjian 2020-2021

出版发行：东南大学出版社
社　　址：南京市四牌楼2号
网　　址：http://www.seupress.com
邮　　箱：press@seupress.com
邮　　编：210096
电　　话：025-83793330
经　　销：全国各地新华书店
印　　刷：南京新世纪联盟印务有限公司
开　　本：889 mm×1194 mm　1/20
印　　张：9
字　　数：352千
版　　次：2022年6月第1版
印　　次：2022年6月第1次印刷
书　　号：ISBN 978-7-5766-0142-8
定　　价：78.00元

本社图书若有印装质量问题，请直接与营销部联系。电话：025-83791830